Series in Display Science and Technology

Series editors

Karlheinz Blankenbach, FH für Gestaltung, Technik, Hochschule Pforzheim, Pforzheim, Germany

Fang-Chen Luo, Hsinchu Science Park, AU Optronics, Hsinchu, Taiwan

Barry Blundell, Waiheke Island, New Zealand

Robert Earl Patterson, Human Analyst Augmentation Branch, Air Force Research Laboratory, Wright-Patterson AFB, OH, USA

Jin-Seong Park, Division of Materials Science and Engineering, Hanyang University, Seoul, Korea (Republic of)

The *Series in Display Science and Technology* provides a forum for research monographs and professional titles in the displays area, covering subjects including the following:

- optics, vision, color science and human factors relevant to display performance
- electronic imaging, image storage and manipulation
- display driving and power systems
- display materials and processing (substrates, TFTs, transparent conductors)
- flexible, bendable, foldable and rollable displays
- LCDs (fundamentals, materials, devices, fabrication)
- emissive displays including OLEDs
- low power and reflective displays (e-paper)
- 3D display technologies
- mobile displays, projection displays and headworn technologies
- display metrology, standards, characterisation
- display interaction, touchscreens and haptics
- energy usage, recycling and green issues

More information about this series at http://www.springer.com/series/15379

Michael E. Miller

Color in Electronic Display Systems

Advantages of Multi-primary Displays

Springer

Michael E. Miller ⓘ
Systems Engineering and Management
Air Force Institute of Technology
Dayton, OH, USA

ISSN 2509-5900　　　　　　ISSN 2509-5919　(electronic)
Series in Display Science and Technology
ISBN 978-3-030-02833-6　　ISBN 978-3-030-02834-3　(eBook)
https://doi.org/10.1007/978-3-030-02834-3

Library of Congress Control Number: 2018959273

© Springer Nature Switzerland AG 2019
This work is subject to copyright. All rights are reserved by the Publisher, whether the whole or part of the material is concerned, specifically the rights of translation, reprinting, reuse of illustrations, recitation, broadcasting, reproduction on microfilms or in any other physical way, and transmission or information storage and retrieval, electronic adaptation, computer software, or by similar or dissimilar methodology now known or hereafter developed.
The use of general descriptive names, registered names, trademarks, service marks, etc. in this publication does not imply, even in the absence of a specific statement, that such names are exempt from the relevant protective laws and regulations and therefore free for general use.
The publisher, the authors and the editors are safe to assume that the advice and information in this book are believed to be true and accurate at the date of publication. Neither the publisher nor the authors or the editors give a warranty, express or implied, with respect to the material contained herein or for any errors or omissions that may have been made. The publisher remains neutral with regard to jurisdictional claims in published maps and institutional affiliations.

This Springer imprint is published by the registered company Springer Nature Switzerland AG
The registered company address is: Gewerbestrasse 11, 6330 Cham, Switzerland

I dedicate this book to my advisors, mentors and family without whose encouragement I would not have gained the experience and knowledge to develop this manuscript.

I would especially like to thank Dr. Helmut Zwahlen who began calling me Dr. Miller many years before I earned the title. His encouragement and work ethic changed my life. I would also like to thank my former colleagues at Eastman Kodak Company. Without their openness to new ideas and their desire to see OLED change the world, I would not have been able to explore this technology. I would especially like to thank my family for their encouragement and patience. Without the support of my wife Karen, son Nathan, and loving parents, none of this work would have been possible. Finally, I would like to thank the late Dr. Lou Silverstein whose work motivated my interest in this field and whose mentorship and friendship have been missed since his untimely passing.

Preface

Color is part of our everyday experience in both our natural and our virtual worlds. Color affects our mood, alertness level, and our ability to recognize and identify objects. We frequently discuss color by making statements such as "the car is blue." However, color is a perception, not an actual characteristic of an object; therefore, it is more accurate to say "I perceive the car to be blue." This subtle difference is important as we think about and discuss color. Color in the natural world is influenced by the characteristics of an object, the light in the environment in which the object is viewed and our own perceptual abilities. Due to its complexity and importance, color has been studied as a science since the nineteenth century and several texts describing color science and its application have been written since.

The body of color science influences the production of dyes and pigments which form the paints and colorants in the products we purchase; the design of lights we use to illuminate our homes and businesses; the cameras we use to capture natural images; the encoding, transmission, and decoding schemes we use to store and transmit images; and display devices we use to view images. In our virtual world, each of these elements influences the color we perceive from an electronic display. However, the influence of each of these elements in the color system has received limited attention from display designers.

In the middle of the 1980s, as an undergraduate student, I was fortunate enough to read a paper by the late Dr. Lou Silverstein. Dr. Silverstein described the potential utility of including a pixel having a yellow color filter to augment the red, green, and blue filtered pixels in a liquid crystal display, commonly referred to as a LCD. Reading this first paper on multi-primary display technology, I came to the realization that I needed to understand color science in much greater detail if I was going to contribute significantly to visual display design.

I began my career working for International Business Machine's visual products division in the late 1980s and early 1990s. I then completed my Ph.D. at Virginia Tech in the Visual Displays laboratory under the guidance of Dr. Harry Snyder and Dr. Robert Beaton.

At the Society for Information Display meeting in the spring of 1995, a former Virginia Tech colleague, in a hushed conversation, pulled what appeared to be a microscope slide from his jacket. He proceeded to attach the slide to a 9V battery. Instantly, the slide lit up to show the word "Kodak" in great, green, glowing, organic light-emitting diode (OLED) letters. I joined Kodak that fall hoping to work on this technology, but, fortuitously, spent several years supporting digital camera development and digital image processing for photofinishing before having the opportunity to join the OLED effort in the spring of 2002. By the end of 2002, we were constructing OLED prototypes with filtered red, green, and blue pixels together with unfiltered white pixels, building upon the knowledge amassed since reading the paper by Dr. Silverstein. This technology was transferred from Kodak to LG Display in 2009 and has since been commercialized in LG's RGBW OLED televisions.

During this journey, I came to understand that many of the assumptions we made about color in display devices were predicated upon assumptions about the performance of cameras or other image capture devices, as well as assumptions about human perception. Further, the design of cameras was based upon assumptions about the objects that were captured and the lighting of the environment in which these objects lie, as well as assumptions about human perception. This led to the realization that to understand color and the design of image capture, transmission, and display devices really requires much more of a systems view of color.

As I have come to understand these assumptions and some common fallacies, it has become evident that to significantly improve these systems, we need a deeper understanding of not only a part of the system, but the entire system. This understanding is particularly important at the current time when technologies employed across many portions of this system are undergoing rapid evolution. For example, light-emitting diode (LED) technology is replacing traditional lighting systems. High dynamic range, light field, and range capture systems are replacing traditional image capture systems. Finally, the introduction of both organic and inorganic LED display technologies, often in virtual or augmented reality systems, is replacing and augmenting traditional liquid crystal displays. Additionally, displays are being manufactured with higher resolution, making multi-primary displays feasible and desirable. Unfortunately, few texts have attempted to capture the system interactions that allow a practitioner in one field to understand the interdependencies of their technology with other technologies in the system.

Not only is technology undergoing rapid evolution but our needs for color are also changing. As it has not been possible to represent color reliably in most of our day-to-day systems, we have settled with the ability to render color to form "natural-looking" or "preferred" images. However, as we move toward virtual commerce, telemedicine, and other applications where the quality of color rendering might influence the success or failure of a virtual retailer or a life-altering medical diagnosis, the need to improve the exactness of color rendering becomes critical to product success.

Preface

The goal of this book is not to make you a color scientist. Instead, my goal is to introduce you to the intricacies of typical imaging pathways which influence display design and the perception of the users of a display. This understanding is particularly important to develop an appreciation for and an understanding of displays having more than three colors of light-emitting elements, displays we refer to as multi-primary displays. We will, therefore, look at some of the attributes of each element in the system before discussing displays. We will then review the technology to enable color in traditional three-primary (RGB) liquid crystal and OLED displays. This discussion will lead us to explore advantages of multi-primary systems. Finally, we will summarize by revisiting the effects of this system on the future of virtual and augmented reality displays. I hope you enjoy this foray into color within the digital imaging system as much as I have enjoyed gathering the knowledge necessary to develop this text.

Dayton, USA   Michael E. Miller

Contents

1 Color from a Systems Point of View 1
  1.1 Natural Color Systems 1
  1.2 Digital Color Systems 3
  1.3 Three Dimensions of Color 4
  1.4 Color in Action 7
  1.5 Perceiving the Importance of Color 9
  1.6 Summary and Questions for Reflection 10
  References 11

2 Human Perception of Color 13
  2.1 Structure of the Eye 13
  2.2 Sensors of the Eye 14
  2.3 Processing in the Eye 16
  2.4 Processing Beyond the Eye 19
  2.5 Defining Luminance and Contrast 21
  2.6 Defining Chromaticity 25
    2.6.1 Tri-stimulus Values 25
    2.6.2 Chromaticity Coordinates and Diagram 27
  2.7 Uniform Color Spaces 28
    2.7.1 CIE 1976 Uniform Chromaticity Diagrams 29
    2.7.2 1976 CIE $L^*u^*v^*$ Uniform Color Space 31
    2.7.3 1976 $L^*a^*b^*$ Uniform Color Space 32
  2.8 General Color Rendering Index 32
  2.9 Defining Human Needs for Color 33
  2.10 Summary and Questions to Consider 35
  References 36

3 Scenes and Lighting 39
  3.1 Measurement of Light 39
    3.1.1 Luminance Measurement 39

		3.1.2	Illuminance Measurement	40
		3.1.3	Power, Efficiency, and Efficacy	41
	3.2	Daylight		42
	3.3	Characteristics of Artificial Lighting		45
		3.3.1	Incandescent	46
		3.3.2	Fluorescent	47
		3.3.3	LED	49
	3.4	Reflectance of Natural Objects		57
		3.4.1	Color in Our Natural World	58
		3.4.2	Relating Color Saturation and Reflected Power	58
		3.4.3	Color Occurrence in Images	62
	3.5	Summary and Questions for Reflection		65
	References			65
4	**Capture, Storage and Transmission Systems**			**67**
	4.1	Digital Capture Systems		67
		4.1.1	Digital Camera Structure	67
		4.1.2	Color Filter Design and Image Reconstruction	69
		4.1.3	Image Integration	72
	4.2	Image Encoding		73
	4.3	High Dynamic Range		76
	4.4	Capturing the Third Dimension		78
		4.4.1	Stereoscopic Image Capture	78
		4.4.2	Multi-view Image Capture	81
		4.4.3	Depth Capture	82
	4.5	Summary and Questions for Thought		83
	References			84
5	**LCD Display Technology**			**87**
	5.1	Display Technology Overview		87
	5.2	Liquid Crystal Display (LCD) Technology Introduction		88
	5.3	Technology Overview		90
	5.4	Contrast and Tone Scale		94
	5.5	Viewing Angle		95
	5.6	Response Time		97
	5.7	LCD Innovations		97
		5.7.1	Backlight Innovations	98
		5.7.2	Quantum Dot LCD	99
		5.7.3	HDR LCD	100
		5.7.4	Temporally Modulated LCD	102
	5.8	Summary and Questions for Thought		104
	References			105

6	**LED Display Technologies**		107
	6.1 Organic Light-Emitting Diode (OLED) Displays		107
		6.1.1 Technology Overview	108
		6.1.2 Electric Properties	110
		6.1.3 Color Performance	113
		6.1.4 Spatial Distribution and Reflectance	119
		6.1.5 Lifetime	121
		6.1.6 Display Structures	123
	6.2 Color Patterning		123
	6.3 White OLED		124
		6.3.1 Other Power Considerations	126
		6.3.2 OLED Summary	129
	6.4 Inorganic Light Emitting Diode Displays		130
	6.5 Summary and Questions for Thought		131
	References		132
7	**Display Signal Processing**		135
	7.1 RGB Display Rendering		135
		7.1.1 Selection of Rendered White Point	136
		7.1.2 Linearization and Decoding	136
		7.1.3 Adapting the Image Data to an Alternate White Point	138
		7.1.4 Target Display Definition	139
		7.1.5 RGB Rendering	141
		7.1.6 Tonescale Rendering	143
		7.1.7 Summary	144
	7.2 Color Rendering for Multi-primary Displays		145
		7.2.1 Simplifying Assumptions	145
		7.2.2 The Color of W Is the Display White Point	146
		7.2.3 Relaxing the White Assumption	148
		7.2.4 Removing the White Assumption	150
		7.2.5 Assuming More Than One Additional Light-Emitting Element	151
		7.2.6 Relaxing the Color Accuracy Assumption	152
		7.2.7 Removing the Stationary White Point Assumption	154
		7.2.8 Summary of Multi-primary Image Processing	158
	7.3 High Dynamic Range		158
	7.4 Summary and Questions for Reflection		159
	References		160
8	**Spatial Attributes of Multi-primary Displays**		163
	8.1 Replacing RGB Elements in Multi-primary Displays		163
	8.2 Resolving Power of the Human Eye		164
	8.3 Display Size, Addressability and Viewing Distance Effects		170

	8.4	Spatial Signal Processing for Multi-primary Displays	172
	8.5	Evaluation of Pixel Arrangements	174
		8.5.1 Display Simulators	175
		8.5.2 Visible Difference Predictors	178
		8.5.3 MTF-Based Evaluation Methods	179
	8.6	Summary and Questions for Thought	180
		References	181
9	**The Multi-primary Advantage**	183
	9.1	An Overview	183
	9.2	Power and Lifetime Effects	184
		9.2.1 RGBW for Filtered White Emitters	184
		9.2.2 RGBW for Unfiltered Emitters	190
		9.2.3 RGBY for Filtered Emitters	192
		9.2.4 RGBYC for Filtered Emitters	194
	9.3	Observer Metamerism	197
	9.4	Modifying Stimulation of the ipRGCs	200
	9.5	Enabling Display Configurations	203
	9.6	Summary and Questions for Thought	207
		References	208
10	**Virtual and Augmented Reality Displays**	211
	10.1	Defining Virtual and Augmented Reality Displays	211
	10.2	Multi-primary Technology in VR/AR Displays	214
		10.2.1 Pixel Arrangements	214
		10.2.2 High Dynamic Range for AR Displays	216
		10.2.3 Dynamic Luminance for VR Displays	218
	10.3	Barriers to VR and AR Display Adoption	221
		10.3.1 Vergence and Accommodation	221
		10.3.2 Overcoming Distortion	223
		10.3.3 Foveated Imaging	224
		10.3.4 Cyber Sickness	225
	10.4	Summary and Questions for Thought	226
		References	227
11	**Multi-primary Displays, Future or Failure?**	229
	11.1	Reviewing the System	230
	11.2	Challenges to Multi-primary Displays	232
	11.3	Supporting Trends	234
		References	237
Glossary		239
Index		245

Abbreviations

8K	A display providing 7680 × 4096 addressable pixels
AR	Augmented reality
a-Si	Amorphous silicon
c.	Circa, meaning "around," "about," or "approximately"
cd	Candela
CIE	International Commission on Illumination
CRT	Cathode ray tube
D65	Standard daylight with a color temperature of 6500 K
e.g.	Exmpli gratia, meaning "for example"
HDTV	A display providing 1920 × 1024 addressable pixels
i.e.	Id est, meaning "in other words"
iLED	Inorganic light-emitting diode
ipRGCs	Intrinsically photosensitive retinal ganglion cells
IPS	In-plane switching
K	Kelvin
L, M, S	Long, medium, and short wavelength cones
LCD	Liquid crystal display
LED	Light-emitting diode
lm	Lumen
LTPS	Low-temperature polysilicon
lx	Lux
m	Meter
M	Modulation (Michelson contrast)
mA	Milli-ampere
ms	Milliseconds
nm	Nanometers
OLED	Organic light-emitting diode
pMAT	Primary matrix
Quad HD	A display having 3840 × 2014 addressable pixels
RGBW	A multi-primary display having red, green, blue, and white elements

RGBX	A multi-primary display having red, green, blue and an additional primary, such as cyan or yellow
S-CIELAB	A spatial filtered version of the CIEs $\Delta E(L^*a^*b^*)$ metric
Sr	Steradian
sRGB	The image standard RGB image encoding standard
TFT	Thin-film transistor
TN	Twisted pneumatic
u', v'	CIE uniform chromaticity coordinates
$V(\lambda)$	Photopic sensitivity function
$V'(\lambda)$	Scotopic sensitivity function
VR	Virtual reality
x, y	CIE chromaticity coordinates
X, Y, Z	CIE tristimulus values

Chapter 1
Color from a Systems Point of View

1.1 Natural Color Systems

Our natural viewing environment is visually complex. Think of a short glance around a wooded landscape. We might see a babbling brook flowing through a wooded area. We see the blue water, gray and white rocks within the water, black shadows, brown tree trunks, brown and green ground cover, and green tree leaves. Imagining this complex and vibrant scene, we perceive these variously colored objects where these changes in color help us differentiate each object from its neighbors. If we look deeper, we see even more detail. We see light reflected from shiny leaves or some of the surfaces of the water. These reflections are very bright and appear white, regardless of the color of the object, from which the light is reflected. We see the rocks at the bottom of the water as the water transmits much of the light, permitting this light to be reflected from the rocks beneath. Each of these complex objects reflect light from the sun in complex ways. A portion of this reflected light then reaches our eyes, which generate electrical signals. These electrical signals are transmitted to our brain, eventually resulting in perceived color.

> We perceive color from complex interactions of light from light sources, which reflect from objects in our environment and are transmitted to our eyes and interpreted by our eye-brain system.

Reviewing our discussion of the natural scene, we see that regardless of the environment, there are at least four influences on our perception of color as shown in Fig. 1.1. These include the sun or other light source. This source provides energy at various wavelengths and typically with some direction. The light from this source is transmitted through some medium, such as the atmosphere or water in the brook before encountering an object. This medium often absorbs a portion of the energy and energy might be scattered by particles such as dust or sand which are suspended

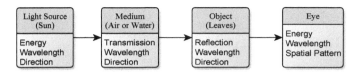

Fig. 1.1 Objects and their attributes in a natural environment influencing the perception of color. Each of these can be viewed as a system component that affects our perception of color

in the medium. The remaining energy then encounters an object, like a leaf, in our environment. This object absorbs some of the light and reflects some of the light at each wavelength. The reflected light also has a direction. Finally, a proportion of the reflected light encounters our eye-brain system. Our eye-brain system then creates the perception of color from the reflected light.

> We don't perceive the color of an individual object, instead we perceive the color of an object relative to all other objects in our environment.

It is important that the light from any single object, such as the leaf in our environment, is not viewed by itself. Instead, light from many objects are typically viewed simultaneously. It is the combination of the light from all of these objects and their arrangement which forms our perception of color, not the light from any single object alone. Additionally, each of these objects can undergo motion, which changes the direction that light is reflected from each object. This direction influences the light entering our eye and our perception of the color of objects in our environment.

Although the system components involved in our perception of color within our natural environment are relatively few in number, multiple attributes of each of these system components influence our perception as indicated in Fig. 1.1. More specifically, we need to consider the spectral emission, transmission and reflection of each object, as well as the direction of this emission, transmission or reflection. Further, the size and any motion of the object can influence the perception of color. Finally, we need to consider not only these attributes of a single object within a complex scene, but a collection of objects within the complex scene.

The number of system components which influence our perception of color increase in digital imaging systems, as each component and processing unit has the potential to influence our perception of color in the displayed image. It is, therefore, useful to simply enumerate some of the key influences in traditional color digital imaging systems.

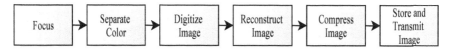

Fig. 1.2 Illustration of the process steps in a digital capture sequence

1.2 Digital Color Systems

When we apply digital imaging systems to capture and convey images of our natural world, these systems have the potential to dramatically affect the colors we perceive. In these systems, an image capture device is used to record and store a representation of the real world. This representation is then rendered onto a digital display. It is this rendered image that we view to perceive the colors in the captured real-world scenes.

In traditional image capture devices (i.e., cameras) light from the natural world is imaged through a lens with a single focal plane. Only the objects in the focal plane of the lens are registered on the sensor with high resolution while the remaining objects in the natural scene are blurred. The sensor will typically have some type of color filter array, which decomposes the image into three or more unique representations of the scene. Design of this color filter array affects the amount of light at each wavelength that reaches each light-receiving element on the sensor. These light-receiving elements then have some native response function and noise level, which affect the image as it is digitized. These attributes of the device are particularly important in darker scenes or darker areas of a scene. Imaging software in the camera then analyzes these digital signals to reconstruct a representation of the scene. Finally, this representation is compressed for storage and transmission. Each of these steps are depicted in Fig. 1.2. At some point, the compressed image is sent to an electronic display, which converts the image to drive signals to customize the presentation of the image to the display. Finally, the image is displayed to be viewed by our eyes.

> Each digital imaging step (i.e., capture, transmission, display) and their associated image processing steps affect the spectral composition of objects, and our perception of the color the objects represented in the digital image.

Each step in this process changes the intensity of the signal provided to each light-emitting element and the spatial information that is eventually displayed. We then view this image with our eyes and perceive the color we associate with each object in the original scene. Color interchange standards have been adopted to provide enough standardization that this process can result in the perception of color in the displayed image which provides a reasonable representation of the color we would have perceived if we viewed the original scene without the intervening digital capture, transmission and display components. However, most manufacturers of consumer and commercial devices maintain highly proprietary signal processing software, permitting differentiation of the devices (e.g., cameras and displays) within this system. While this complicates the predictability of the final color provided by the system, it also motivates competition to improve the processing steps along the way. We will dedicate much of this book to expanding this discussion; however, it is important to understand that each of these steps can have a significant influence on the color we perceive.

1.3 Three Dimensions of Color

As we consider color further, it is important to realize that we are typically taught to think about color in two dimensions. For instance, as children in art class we may learn about additive and subtractive color. In a subtractive system, such as paints or printer inks, when we add yellow and magenta, we are left with red. This is often depicted through a diagram such as a color wheel as shown in Fig. 1.3. As shown in this wheel, we begin with the 3 secondary colors. These include yellow (shown at the top of the figure), cyan (shown in the bottom left), and magenta (shown in the bottom right). If we add a little bit of magenta paint or ink to yellow, the color begins to appear a little more reddish as the yellow and magenta ink both permit red to pass through them while yellow absorbs blue and magenta absorbs green light. Adding more magenta makes the color a darker red. When we mix yellow, magenta and cyan, as shown at the center of this figure, the result is to subtract all of the light before it is reflected from the paper and we are left with black. Therefore in this system as we mix these paints or inks we are selecting the amounts and wavelengths of light to be absorbed so that they cannot be reflected from the surface on which they are painted, such as a white sheet of paper. The wheel depicts the transition of color between the three secondary colors to obtain a full palette of colors.

In electronic displays, we typically utilize additive rather than subtractive color. That is we begin with three primary colors of light, typically red, green and blue.

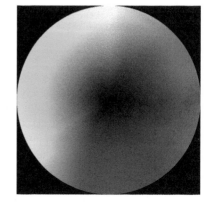

Fig. 1.3 Illustration of subtractive color wheel. Different amounts of magenta, yellow and cyan are added to absorb light before it is reflected from a surface. Varying amounts of these dyes can be used to form primary colors and when all 3 colors are overlaid, all light is absorbed, resulting in black

1.3 Three Dimensions of Color

Fig. 1.4 Illustration of additive color wheel. In this diagram, different amounts of red, green, and blue are added to form a full range of secondary colors between each primary and white at the center of the circle

As we add differing amounts of these three colors of light, we form various colors. In an additive system, the color wheel can be modified so that we begin with certain amounts of red, green, and blue light, such as shown in the color wheel in Fig. 1.4. In this figure, green light is depicted at the top, blue near the bottom left corner and red near the bottom right corner. By adding two or more colors of this light together, we obtain other colors of light. Specifically, the secondary colors (i.e., cyan, magenta, and yellow) are formed by adding light from pairs of these primaries and white is formed in the center of the circle by adding light from all three primary colors.

This same thought process is carried into academic discussions of color. When exploring color science, we are typically first introduced to color in terms of the International Commission on Illumination (CIE) 1931 x, y chromaticity space, as represented in the 1931 CIE chromaticity diagram shown in Fig. 1.5 [2]. As shown, this color space is a two-dimensional (x, y) space. It includes a horseshoe-shaped area which represents all colors we perceive as a human. The horseshoe-shaped portion of this boundary represents highly saturated light, such as the light provided by a laser. The line at the bottom of the horseshoe simply connects the most saturated, yet visible, blue to the most saturated, yet visible, red. As we move towards the center of this color space, colors become less saturated. These colors are achieved by mixing together multiple wavelengths of light. In fact, the point in this diagram represented by coordinates at $x = 0.33$ and $y = 0.33$ is referred to as equal energy white. This color can be achieved by mixing equal amounts energy at every wavelength across the range of visible wavelengths. In reality, this color space is a transformation of the additive color wheel with blue at the bottom left, red at the bottom right and green at the top. Magenta colors lie along the line at the bottom, yellow colors lie between red and green and cyan colors lie between green and blue.

In most electronic displays, we can plot the chromaticity coordinates of each colored light emitting element, such as the red, green, and blue squares shown in Fig. 1.5. In an electronic display we then form color by adding together the light from these three light-emitting elements to form less saturated colors. As a result, we are able to form the subset of colors enclosed in the triangle formed by joining

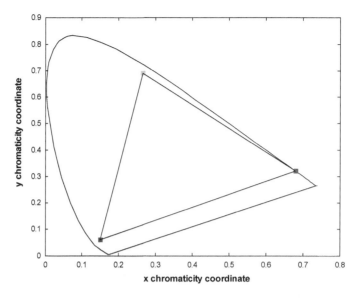

Fig. 1.5 CIE 1931 chromaticity diagram containing coordinates for red, green, and blue display pixels. Also shown is the triangularly-shaped display gamut, which represents the range of colors a display having these red, green, and blue emitters can produce

the chromaticity coordinates for the three light-emitting elements (primaries). For example, by adding together varying amounts of light from the red and green primaries we can produce colors that lie on the line joining these two primaries. The triangle is typically called the display gamut and encompasses all colors that the display is capable of producing.

> We perceive color as a function of relative luminance in our environment. To specify color, we must specify at least 2 chromaticity dimensions and one dimension of relative luminance.

It is important to recognize that while color is often plotted or discussed in a two-dimensional space, this discussion can be problematic. For example, looking back at Figs. 1.3 and 1.4 we see that some colors are missing. Where is the brown of the tree trunks in our example scene? This color is simply not present in either figure. How can this be? If this diagram represents all colors, then brown must be present. We can ask a similar question for our black shadows or gray rocks from our stream when we look at Fig. 1.4. The fact is that the chromaticity coordinates of brown, black and gray are shown; however, these colors are formed with different relative luminance values than the colors shown.

The fact that the relative luminance of objects within a scene has a strong influence on our perception of color is an important concept in color perception. Further, this concept clearly illustrates that color is a perception or an impression created within our eye-brain system and is not a physical attribute of objects in the world. Relative luminance refers to the perceived amount of light (i.e., luminance) that enters our eye from an object as compared to the current state of adaptation of the human eye. In our natural scene, when we are standing in the open field in the sun, our eye is adapted to the large amount of light available from direct sunlight. However, after we walk into the trees our eye adapts to the lower amount of light that is present in the shade of the trees. This adaptation state of our visual system can permit us to perceive an object as orange when the amount of light reflected from the object is high relative to the amount of light to which our eye is adapted (i.e., when the object has a large relative luminance). However, if we were to take that same object and place it in a much brighter environment so that our eye is adapted to the much higher amount of light without permitting the light from the object we perceived as orange before to change, the object would appear darker compared to other colors in the scene (i.e., have a lower relative luminance) and we would perceive the color of the object to be brown.

It is easy to be confused by this discussion. It is possible to interpret this discussion to imply that if we were to look at an object we perceived as orange in a dim environment and increased the amount of light in the environment, we would suddenly see the object as brown. In reality most natural objects reflect light from a light source like the sun. Therefore, as we move an object from a dimly lit environment to a brightly lit environment, the object simply reflects more light. Therefore, in natural environments the relative luminance of objects remains constant regardless of how much light is in an environment. As a result we do not perceive changes in color as the amount of light in an environment changes, at least when the adaptation state of our eye changes as the amount of light in the environment changes.

Therefore, understanding relative luminance we now can say that brown has color coordinates similar to orange or yellow, it is just darker. Similarly, gray and black are darker versions of white. That is gray and black have lower relative intensity than white. For this reason, it becomes important not to discuss color in only two dimensions. To fully understand color, we must consider color in three dimensions, where the third dimension represents either luminance or relative intensity. As we will see, the need to consider color in this third dimension is particularly important if we are to understand the advantages of multi-primary displays. As we have illustrated in this section, it is also important to realize that objects do not have color, they simply reflect or emit various amounts of energy at each wavelength. It is our visual system's interpretation of this energy that permits us to perceive the color of each object.

1.4 Color in Action

While we are introducing the discussion of color, we must ask: "Is three-dimensions adequate to represent all colors?" Returning to our discussion of the natural landscape, we remember that the scene changes with time. The water flows over the rocks, the wind moves the leaves, a cloud passes between us and the sun or we walk out of the direct sun into the shade of the trees. With the passage of time, the lighting in our environment changes. Perhaps most important among the previous list of influences is the passage of the clouds over the sun or our walk into the shade. With each of these changes, the amount of light entering our eye is reduced. Areas of the scene that looked like dark shadows slowly transform into colorful areas of our scene as our eye adapts to the lower luminance. Suddenly we can see color where we could not before.

At the same time we begin to be able to see into the darker regions of the wooded area, other changes occur as well. As our eye adapts to the reduced light, not only do the dark areas of our scene begin to show color, but other colors change as well, typically becoming more colorful. Simultaneously, if we quickly glance back out into the open area after walking into the shade, everything in the open area appears brighter and less colorful.

> In the natural environment, light is not constant but undergoes change and our eyes adapt constantly to adjust to these changes.

Through this example, we can come to three understandings. First of all, the leaves in the shadows are not themselves green, but we perceive them as green once we walk into the shade or the cloud covers the sun. Remember, we perceived them as black before we took those few steps or when the sun was providing its beautiful, bright light. Is the leaf always green? It is reflecting light similarly, but our perception has certainly changed. Secondly, our perception of color can change with time if the lighting conditions and the objects in the environment change. Color does not necessarily change with time, but it does change over time if the lighting conditions change over time. As we will see, this concept can also be important when we talk about multi-primary displays.

Finally, it is worth pointing out that as we walked into the shade or the cloud passed over the sun: the amount of light available to our eye did not change by a few percent. Instead it was reduced dramatically. In fact, the amount of light available to our eye was likely much less than one tenth of what it was when we were in full sun. During this period of variation we saw a fairly dramatic transformation of our visual world as the intensity and wavelengths of light reflected from objects in our environment changed and many of our visual mechanisms responded to this dramatic change by producing changes in the colors we perceived. Similar changes occur again when we walk from the shade into the bright sunlight, likely requiring us to squint, place our hand over our eyes or take other conscious action to reduce the amount of light entering our eye until our eye is able to adapt to the higher light level. We are often consciously aware of these dramatic changes and they influence our beliefs about our surroundings. In current electronic displays, which have the ability to change their luminance over a relatively small range, such dramatic changes are difficult to produce and our ability to perceive changes on our displays as realistic are often limited by this artifact.

1.5 Perceiving the Importance of Color

Considering the fact that our perception of the color of an object can change over time, one might ask "Is it important to carefully control our digital displays in an attempt to deliver precise color?" After all, if the perceived colors of objects in our natural environment are constantly changing, will users of displays appreciate the result produced by this effort or are we better off spending precious development dollars and effort in other areas to improve the display viewing experience?

In fact, color is critical to the human experience. The presence of highly saturated colors attracts our attention, guiding our eyes around a scene on the display and helping us to quickly locate high interest objects having highly-saturated color [1]. Color is very useful in helping us to differentiate one object from another in complex scenes. Further the changes in wavelength of light improve our ability to identify particular objects as unique (e.g., finding and recognizing a berry as a ripe berry would be much more difficult if the berry was green as compared to red or purple). Therefore, displays with well-designed color are more likely to appear bright and vibrant to an end user. The wavelengths and intensity of the light produced by a display can be manipulated to increase the perceived difference between displayed colors, thus increasing the vibrancy of colors. However, improper enhancement can quickly lead to garish-looking colors as there are ranges over which we expect some colors to change and changes beyond those boundaries are unacceptable to many users.

> High fidelity color rendering is important in today's electronic displays and even more important in tomorrow's electronic displays.

The role of the electronic display is undergoing change. Traditionally, these displays have simply been used for entertainment or to support office chores. However, as we enter a virtual society, purchases are increasingly made on line. Decisions about clothing or accessory purchases are now being made through electronic media and the purchase of objects having a different color than the color of the object on an electronic display is likely to result in dissatisfied customers and increased product returns. In fact, online retailers constantly struggle with naming the color of items to attempt to reduce the number of returns and yet these returns play a significant role in their business model.

This virtualization of society goes beyond simple online purchases. Applications like virtual medicine have the potential to lower the cost of health care. However, in certain diagnostic areas, improper color rendering might lead to difficulties in diagnosis. Imagine a dermatologist attempting to diagnose the state of a blemish on the skin. Is it possible that the shape of the outline of this blemish might appear different on one display than another based simply on capture and display conditions? Certainly the intensity and wavelengths of light reflected from this blemish may change across its width. With improper rendering, the intensity and wavelength of light near the edges of the blemish may become more like the surrounding area than the center of the blemish. Therefore, with changes in color rendering, one may incorrectly perceive regions near the edges of the blemish as part of the surrounding area, rather than as part of the blemish. It is then possible for this change in perception to lead to a change in diagnosis as a result of the perceived change in shape of the blemish. Suddenly rendering of color on a display might have life changing consequences. As such, high-quality color rendering will be even more important in tomorrow's displays than it is in today's displays.

1.6 Summary and Questions for Reflection

In this chapter, we have discussed some of the complexity of color and alluded to the advantage of multi-primary displays in providing consumer advantage through the improvement of three-dimensional color perception. In subsequent chapters we will begin to explore each of the system influences on perceived color, providing insight into the impact of each system component shown in Figs. 1.1 and 1.2, before delving deeper into display color rendering and multi-primary displays. Through this journey, I hope to illustrate the advantages of multi-primary displays. As we will see, among the advantages of multi-primary displays are the following: (1) a reduction in display power consumption and an increase in display peak luminance, especially for displays employing color filters; (2) the ability to improve the consistency of color appearance between individuals (i.e., improved color metamerism); (3) an improved ability to adapt the display to trade brightness for color saturation during time windows when the human eye is adapting; (4) the potential to enable display architectures which are impractical when considering 3 primary displays, and (5) potential changes in rendering to control the effect of the light produced on human biorhythms and alertness. We will discuss each of these advantages in the context of this broader imaging system. Given this background, here are a few questions to consider before moving on to the next chapter.

1. If our perception of color is affected by relative luminance, why is each generation of electronic displays designed to provide higher absolute luminance?
2. We discussed color as having three dimensions, two color dimensions and relative luminance. Then we discussed the fact that luminance changes with time in many environments. Should we include a fourth dimension of time?
3. If we were more confident in the accuracy of color imaging, could we create new, more useful, virtual experiences?

References

1. Christ RE (1975) Review and analysis of color coding research for visual displays. Hum Factors 7(6):542–570
2. CIE (1932) Commission internationale de l'Eclairage proceedings. Cambridge University Press, Cambridge

Chapter 2
Human Perception of Color

2.1 Structure of the Eye

To discuss the eye's function, it is useful to understand the structure of the eye, as is depicted in Fig. 2.1. As shown in this figure, light enters the eye through the pupil. The area of the pupil area is covered by a protective layer referred to as the cornea and the size of the pupil is controlled by the iris. The iris dilates to increase the size of the pupil and contracts under control of a sphincter muscle to decrease the size of the pupil.

It is often misunderstood that the primary mechanism the human visual system uses to adapt to changes in light level is through changes in pupil size. It is true that the human eye/brain system adjusts the size of the pupil in response to changes in light level, constricting the pupil to permit less light into the eye as the amount of light in the environment increases and expanding the pupil to permit more light to enter the eye when the amount of light in the environment decreases [26]. However, pupil size responds to other factors. For example, pupil size changes in response to

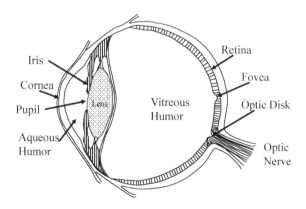

Fig. 2.1 Anatomical structure of the human eye. Figure adapted from Coren and colleagues [5]

© Springer Nature Switzerland AG 2019
M. E. Miller, *Color in Electronic Display Systems*, Series in Display Science and Technology, https://doi.org/10.1007/978-3-030-02834-3_2

the accommodation state (c. focus) of the eye [9], as well as, changes in emotion [24]. For example, the size of the pupil is known to increase with increased arousal as well as increases in workload [19]. Further the size of the pupil can only change the light level inside the eye by about a factor of 10 while the human visual system is capable of adapting to light levels which change by a factor as large as 10^6. The pupil does, however, respond rapidly to changes in light level, permitting other adaptation mechanisms within the human eye-brain system to respond less rapidly.

The lens of the eye focuses the light from a plane in the world onto a sensor plane, referred to as the retina, at the back of the eye. The process of changing the focus plane of the lens is referred to as accommodation. Like any lens, the lens of the eye has a finite depth of field. As a result, only objects near the plane where the lens is focused will be focused on the retina and objects will be blurred which are distant from the eye's focal plane. Therefore, the plane of focus of this lens must change frequently as we look at different objects in the world. Within the lens of the eye, focus is changed as ciliary muscles contract, causing the lens to thicken and permitting it to focus near objects.

It is useful to note that there is a significant interaction between pupil size and the function of the lens. As pupil size decreases, the depth of field of the lens increases. This increase in depth of field reduces the need for exact accommodation to specific objects in our environment. With age, the ability of our eye to accommodate is reduced as the lens of the eye becomes less flexible. However, pupil size also generally decreases with age [26], potentially providing a method for partially compensating for the reduced accommodative function of the human eye.

As light reaches the retina, it first encounters retinal ganglion cells, which collect electrical pulses produced by the light-sensitive cells in the retina. These retinal ganglion cells converge at a location on the retina forming the optic nerve, which connects the eye to the brain. Beyond the retinal ganglion cells are the sensors of the eye. Finally, the back of the eye is formed from a pigmented layer, which commonly absorbs any of the light which pass the retinal ganglion cells and the photosensitive sensors within the eye. The fovea is a small area on the retina with the highest density of sensors capable of sensing color. This region, while small compared to the size of the retina, plays a significant role in color perception.

Importantly, the eye is filled with a thick fluid, with the chamber in front of the lens being filled with aqueous humor and the chamber behind the lens filled with vitreous humor. This fluid transports nutrients and waste products between capillaries in the eye and the structures of the eye. This fluid also provides an even pressure on the exterior surfaces of the eye, permitting the eye to take the orb-shape which is necessary for proper function.

2.2 Sensors of the Eye

It is well known that the human eye of individuals with normal color vision contains four sensors which are responsible for gathering light useful in supporting vision.

2.2 Sensors of the Eye

These include very light sensitive rods, which primarily support vision in very low light conditions, as well as three less-sensitive cones, which support vision in daylight conditions. More recently, it has been shown that the eye contains at least one additional sensor which primarily supports functions other than vision [2, 7]. These sensors, referred to as intrinsically photosensitive retinal ganglion cells (ipRGCs), appear to influence our biorhythms and level of alertness [15, 23]. While the three cones are primarily responsible for influencing our perception of color, each of the other types of sensors can, less directly, influence our perception of color under certain conditions.

> Our eye includes at least 5 sensors, 3 cones to support color vision, rods to support vision in low light and ipRGCs to aid the regulation of biorhythms.

It is important to understand the sensitivity of each of the sensors within our eye. Figure 2.2 shows the relative sensitivity of each of the five sensors we have just discussed. The short (S), medium (M) and long (L) wavelength cones are each sensitive to and integrate different wavelengths of light. As shown in the figure, the S cones are sensitive to wavelengths as short as 380 nm and our L cones are sensitive to wavelengths as long as 720 nm. Between these two limits, there is substantial overlap, particularly between the L and M cones, but among all of the cones. That is our color vision system is sensitive to all wavelengths between these boundary values and at least 2, if not all three of our cones will provide a response to most wavelengths of light within this region.

It is also important that the rods are sensitive to a smaller wavelength region, from about 400 nm to around 620 nm with a peak sensitivity near 505 nm. Finally, the intrinsically photosensitive retinal ganglion cells (ipRGCs) are sensitive to wavelengths between about 400 and 600 nm, with a peak sensitivity near 485 nm.

As is well known, the sensors of the eye are distributed unevenly. The vast majority of the cones are located in a small area of the retina, referred to as the fovea. The optics of our eye image a small angle of our world, only about 2° in extent, on our fovea. Therefore, we only see this small portion of our world with high resolution at any moment in time. We must then move our eyes to position this fovea to view other parts of the world with high resolution. These eye movements occur every 0.1–0.7 s to permit our brain to integrate our perception of the world. Within our fovea, there are many fewer cones that are sensitive to short-wavelength energy, typically associated with blue light, than cones which are sensitive to longer wavelengths of light.

Exploring the peripheral region of the retina around the fovea, this region is populated by rods, with significantly fewer cones. The number of rods and cones continue to decrease as the distance from the fovea increases. As such, the resolution of our eye is greatest in the fovea and generally decreases as the distance from the fovea increases [16]. As a result, our eye forms a very high resolution image of the 2° angle in the center of our visual field and the resolution decreases as the angle from the center of our visual field increases. Similarly, our perception of color is

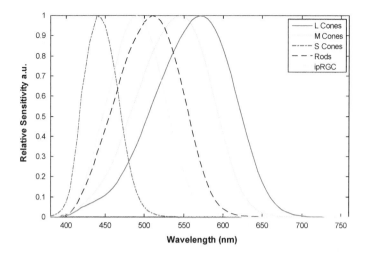

Fig. 2.2 Relative sensitivity of the sensors of the human eye, corrected for absorption by the remaining eye structures [17]. Shown are sensitivities for the long (L), medium (M), and short (S) wavelength cones, rods and intrinsically photosensitive retinal ganglion cells (ipRGC). Curves are plotted from data provided by Lucas et al. [17]

primarily influenced by the sensors in our fovea, although we perceive color across a large portion of our visual field. The resolution of this color signal is much lower outside the foveal region.

2.3 Processing in the Eye

While the light entering our eye is collected by our rods and cones to form a visual image, it is not signals from these individual sensors that are transmitted to our brain to form the perception of color. Instead, signals from individual sensors are gathered by retinal ganglion cells within the eye. Each retinal ganglion cell collects signals from multiple rods and cones. While the specifics of these connections are beyond the scope of our discussion, there are three important emergent properties of these elements which are important to our discussion. First, the signals leaving the eye generally correspond to a luminance and two color (chrominance) signals. Second, the signals are primarily difference signals, which are important in the adaptation of our eye to light. Third, the resolution of these signals varies across our visual field and are different for luminance and each of the chrominance channels. It is important to discuss each of these elements.

2.3 Processing in the Eye

> The signals leaving our eye generally correspond to a luminance and two color difference signals.

Included in the processing that occurs within our retina, a signal can be produced by collecting signals from our rods and all three of our cones. By summing this output, one could form a signal which generally corresponds to the amount of light within the environment. In fact, some basic models of processing in the eye include this element [10]. Importantly, however, the eyes are not necessarily forming a signal corresponding to the amount of light within our environment. Instead the retinal ganglion cells within the eye collect signals from a spatial area within our visual field and provide signals corresponding to the change in luminance across the spatial area that they sample. Therefore, these cells are really responding to changes in light within their collection area, as a function of distance, a function of time, or a combination of these factors.

Similarly, the same basic models of human color vision assume that a second signal is produced from a difference between the L and M cones. This signal then corresponds to a red-green difference signal. Finally, a third signal might be produced which sums the response of the L and M cones, which would result in a yellow signal. However, the same cells responsible for this signal additionally compute the difference between this yellow signal and the output of the S cones, resulting in a blue-yellow difference signal. These difference signals might then correspond to the axes in a color wheel. Similar to the signal for luminance, however, the eyes do not deliver a signal corresponding to the amount of red-green or blue-yellow light, but produce signals which correspond to the changes in these colors of light across an area within the retina. As a result, our eyes produce signals which correspond to the changes in the amount of light or one of these two color signals across a spatial area or as a function of time. As a result, within our visual processing system, most of the information pertaining to the absolute amount of light within our environment is lost.

> The fact that our eyes produce difference signals is an important aspect of vision, permitting our eye to adapt to changes in the amount of light within our environment without our awareness.

Adaptation of our eye is an important process in color vision as we discussed in the first chapter. When looking into the woods from the sunny field, we were unable to see the colors of objects in the trees. Once we enter the trees and our eyes adapt to the lower light levels, the color of objects under the trees suddenly become apparent. Adaptation takes place in our eye through several mechanisms with different time scales of response. These include changes in pupil size, bleaching of the sensors, and changes in gains within our retinal ganglion cells.

Changes in pupil size occur rapidly, but permit the eyes to adapt to only relatively small changes in light level within our environment. While the effect of pupil size on our visual system's response is important, it is among the smaller effects.

The sensors within our eyes undergo chemical changes with changes in light level. For instance, high levels of light over saturate the rods within our eyes, bleaching them, such that they no longer produce changes in signal with small changes in light level. As such these sensors likely do not contribute significantly to our visual experience in bright light conditions. Adaptation of these sensors, particularly to changes from light to dark can require several minutes, with more than a half hour required to adapt from bright daylight conditions to nighttime conditions. This adaptation is slow, but permits our eyes to function over several log changes in light level.

Finally gains can be adjusted within our retinal ganglion cells to permit rapid adaptation of our eyes over a moderate range of light levels. This adjustment is perhaps the most important mechanism present for daylight color perception. Unlike the other two mechanisms, these gains do not necessarily occur over our entire visual field. Instead, this adaptation can occur locally, permitting us to see brightly lit objects within one area of our visual field and darker objects at some other location in our visual field.

It is also important that adaptation generally occurs without our conscious awareness. It is only in the presence of very large changes in light level that we might find ourselves squinting or shielding our eyes to reduce the amount of light entering our eyes. Alternately, we might occasionally become consciously aware of the need to dark adapt when we cannot see detail that we expect to be present. Otherwise, we are generally unaware that the light level around us is changing and we are not aware of the absolute light level in our environment. That is we often do not perceive much, if any, change in light level between midday and evening lighting, although light level could differ by several orders of magnitude.

> The resolution of the signals vary across our visual field and are different for luminance and each of the chrominance channels.

The array of sensors, retinal ganglion cells and their connections within our eye influence the resolution of our eye. Importantly, the spatial resolution of our eye is highest when detecting changes in light level. The resolution of our eye is lower for the red-green signal and even lower for the blue-yellow signal. At the moment, we will leave the detail of these differences vague, returning to them at a later time. It is important to recognize, however, that the resolution of our eye is also influenced by the focus of light on the sensors by the lens of the eye and flare which is produced by spreading and scattering of the light within the human eye. The overall influence of these factors is to reduce the perceived color saturation of small objects and to reduce our ability to discriminate low luminance detail in the vicinity of brighter objects. Each of these effects are important as we discuss methods for rendering to multi-primary displays.

While an absolute measurement of the amount of light might not be present in the imaging system of our eye, the ability to detect particularly high levels of illumination can influence the function of our visual system, including pupil size, which affects adaptation.

Beyond these effects, pupillary response as part of adaptation can significantly influence the depth of field of our eye (i.e., the range of distances which are in focus on the retina) and flare within our eye. Specifically, smaller pupils provide greater depth of field and reduces flare within our eye. While ipRGC response does not directly affect vision, the response of these cells do directly affect pupil dilation. As the ipRGCs and the rods and cones to which they connect are located primarily in the peripheral retina, it is likely that light applied to the peripheral retina is more effective than light applied to the foveal region in driving pupil constriction [22]. Consistent with the sensitivity function of the ipRGC, the pupil constricts more after exposure to high intensity blue wavelengths of light than when exposed to red light having equal photopic luminance [14] and recent research has shown that further decreases in pupillary response can be driven by light which is slowly flickered between blue and red [22].

Besides influencing our pupillary response, the ipRGCs also affect the adjustment of our circadian rhythm through photo-entrainment. For instance signals from ipRGCs have been shown to suppress the release of melatonin, a chemical in the blood associated with drowsiness. It is now believed that the response of these sensors may be responsible for the fact that shift workers exposed to bright light for a few hours during the night shift exhibit improved alertness, cognitive performance, and shifts in circadian clock than similar shift workers performing in relatively low light working conditions [6]. Originally, this effect was believed to be attributable solely to the suppression of melatonin. However, recently, it has been shown that light can improve alertness even during early daylight hours when melatonin release is believed to be minimal [21]. Regardless of the mechanisms, it is clear that bright light, and especially bright light to which the ipRGCs respond can improve alertness, cognitive performance, and circadian cycle entrainment as well as influence pupillary response.

2.4 Processing Beyond the Eye

The cells within the eye play a significant role in converting the energy which enters the human eye to luminance and chrominance signals, which are important to understanding color perception. However, it is important to understand a few of the processes in the human brain as well. Figure 2.3 depicts the visual pathways from the scene to the visual cortex.

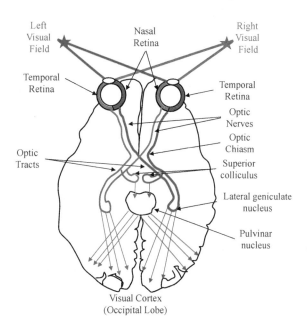

Fig. 2.3 Visual pathways from the scene to the visual cortex. Adapted from Coren and colleagues [5]. Note items associated with the left visual field are represented in blue while items associated with the right visual field are represented in red

As shown in Fig. 2.3, as objects in the left visual field are imaged by the lens of the eye these objects are represented on the right side of the retina in each eye while objects in the right visual field are imaged on the left side of the retina in each eye. As the signals are collected and processed in the eye the signals leave the eye through the optic nerve and pass into the brain. The two optic nerves then join at the optic chiasm. At this junction, nerves from the nasal side of each eye cross to the complimentary side of the brain. As a result, the signals originating from the left visual field pass into the right side of the brain while signals originating from the right visual field pass into the left side of the brain. As the nerves leave the optic chiasm, the resulting nerve bundles are referred to as optic tracts.

Beyond the optic chiasm, a portion of the optic tracks extend to the superior colliculus which is located on the brain stem. The signals then pass to the pulvinar nucleus and the visual cortex. The signals passing along these pathways may be responsible for stimulus or bottom-up driven eye movements and potentially spatial localization of signals.

The remaining signals pass to the lateral geniculate nucleus. This area redistributes functionally different signals to either the parvocellular or magnocellular layers of the visual cortex. Generally, the magnocellular layers provide a fast response while the parvocellular layers provide a slower response. The magnocellular layers appear to be responsible for processing course achromatic (c. luminance) signals and provide a robust response to motion. On the other hand, the parvocellular layers of the visual cortex are responsible for processing fine achromatic signals, providing the high spatial resolution of the visual system and stereoscopic vision. The parvocellular layers also provide color processing, and may provide the sole role in chromatic contrast, and the perception of hue and saturation within images [8].

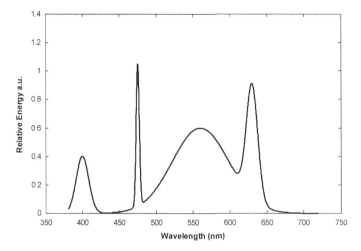

Fig. 2.4 An example complex lighting spectrum, containing multiple emission peaks with varying amplitude and bandwidth

2.5 Defining Luminance and Contrast

Thus far, we have discussed vision in qualitative terms. That is, we have discussed some of the general function and performance, without providing quantitative methods for specifying performance. However, to really understand display design, we need quantitative methods for specifying visual performance. Therefore, the remainder of this chapter describes some initial methods we might use to quantify the performance of the human visual system.

As we begin to talk about quantifying human vision, we have to think a little about what we want to accomplish. One of our basic needs is a method to quantify the amount of light and the color of light for any light spectrum. For instance, we might see a blue LED which emits light at 420 nm and state unequivocally that the LED produces blue light with some power, measured in Watts. What happens when we have a complex spectra, like the one shown in Fig. 2.4? Note that this spectra has multiple emission peaks with different amplitudes and widths. How do we describe the energy from this source in a way that is relevant to the human visual system? Can we tell by looking at this spectra, the color that a human will perceive? Can we tell whether a light with this spectra will appear to a human as brighter than the blue LED?

Early color scientists took the approach that if we could measure the relative sensitivity of the human eye to each wavelength of visible energy, we could multiply the optical power at each wavelength by the eye's sensitivity at that wavelength and then sum this quantity across all wavelengths [27]. If this could be accomplished appropriately, we could then compare this value for any two light sources and the

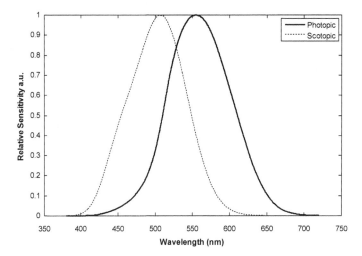

Fig. 2.5 Curves illustrating human vision system sensitivity curves for photopic (daylight) and scotopic (dark) viewing conditions. Curves created from table provided in Wyszecki and Stiles [27]

resulting value should provide information regarding the relative brightness of the two light sources. Appropriately weighted, we typically refer to this metric as luminance.

> Luminance is a quantitative metric which correlates with changes in perceived brightness for objects having the same or different spectral composition.

Of course an issue with this metric is that our eye's sensitivity is not constant. During daylight viewing conditions, or when viewing most bright consumer electronics displays, our eyes are adapted such that their sensitivity is determined by the cones. However, when viewing displays which output very little energy at night, our eyes may be adapted so that the sensitivity of our eyes are determined predominantly by the response of our rods. Therefore, there is not a single sensitivity function for our eyes. Figure 2.5 shows the sensitivity function of the eye during daylight conditions, referred to as the photopic sensitivity function ($V(\lambda)$), as well as the sensitivity function of the eye during low light conditions, referred to as the scotopic sensitivity function ($V'(\lambda)$). Notice, that as we described earlier, $V'(\lambda)$ has a peak near 505 nm, corresponding to the peak sensitivity of the rods. The peak in the $V(\lambda)$ is longer having a peak near 560 nm, corresponding to the peak sensitivity of the combined response of the three cones. It should also be recognized that each of these curves are standard curves that are intended to represent an average human observer. Differences, of course, occur between individuals and our sensitivity changes as we age. Therefore, it is important to recognize that these curves are not exact.

There is one additional issue that we should discuss and that is: "How do we quantify the performance of the eye under conditions that are somewhere between

2.5 Defining Luminance and Contrast

photopic and scotopic?" In this range, referred to as mesopic visual conditions, both the rods and the cones are active and contribute to vision. While this question has received considerable research, the current approach is to apply an average of these two functions where it is assumed that the scotopic function will be applied to any light levels below 0.005 cd/m^2, the photopic function will be applied to any light levels above 5.0 cd/m^2 and an average of these two functions, weighted by the proportion of the distance between these two light levels will be applied for intermediate values [12].

Luminance is then calculated from a spectral power measurement using Eq. 2.1. As shown, this equation computes the radiance $S(\lambda)$, measured in Watts, multiplied by the eye sensitivity function $V(\lambda)$ at each wavelength and sums this value across the visible spectrum. The sum is then normalized by a constant, K_m, which is equal to 683.002 lumens per Watt. Therefore, luminance is reported in units of lumens. A similar calculation is performed for the scotopic illumination conditions, only applying the scotopic sensitivity function $V'(\lambda)$ and a constant of 1700 lumens per Watt.

$$L_V = K_m \sum_{\lambda=380}^{760} (S(\lambda)V(\lambda)) \qquad (2.1)$$

Perceived brightness is not linearly related to luminance, but is better represented by a logarithmic response.

Luminance is an important metric. It allows us to take lights having different spectra and determine which of these lights will be perceived by the user as brighter. By simply computing the luminance of the blue LED and the luminance output of the spectra shown in Fig. 2.4, we determine which will be perceived as brighter by a typical human observer simply by selecting the light with the higher luminance value.

It is important, however, to recognize that our perception of relative brightness is not linearly related to luminance. In fact, the perception of relative differences in luminance changes depending upon the range of luminance values we are considering [25]. However, within the range of luminance that most commercial displays operate, the changes in brightness can be approximated by assuming the ratio of the change in luminance to the absolute luminance is relatively constant. This relationship is often known as a Weber's Law relationship. Under these conditions, small changes in low luminance values appear similar to much larger changes in luminance when the absolute luminance is high. Said another way, the eye's response can be approximated as a logarithm of luminance. Several different models can be used to approximate this relationship. Figure 2.6 shows a model referred to the Digital Imaging and Communications in Medicine or DICOM model [1].

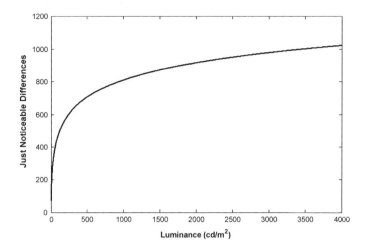

Fig. 2.6 Just noticeable differences as a function of luminance, as computed using the DICOM standard [1]

The model depicted in Fig. 2.6 relates just noticeable differences in luminance to absolute luminance value. As this figure shows, at low luminance values, small changes in luminance are noticeable. However, at high luminance values, much larger changes in luminance are required before the change is noticeable.

As discussed earlier, the retinal ganglion cells determine differences in light level, rather than measuring the light level itself. That is, they respond to this difference or luminance between light and dark, red and green or blue and yellow light. Therefore, we typically characterize our perception in terms of changes in light. It is therefore, important that we define metrics of contrast as well as traditional color metrics.

> Our eye does not determine absolute light level, but it is excellent at detecting changes in light level. That is, it is excellent at seeing the contrast between objects of different light levels.

Unfortunately, an agreed upon, standardized measure of contrast does not exist. Here we will define two measures of contrast which are commonly used and have utility. The first of these measures is the simple contrast ratio. This metric is commonly used within the display industry and expresses the ratio of the luminance of a bright area in the display to the luminance of a dark area on the display. Traditionally, this ratio is used as a metric of display quality, providing the ratio of the highest white luminance a display emits to the darkest black the display emits. Unfortunately, this metric tells us very little about the appearance of the display. In fact, some modern displays can create a black with zero luminance, resulting in a contrast ratio of infinity. This value quickly approaches infinity for very low light conditions and is misleading as changes in very low light level blacks may have little influence on our perception of the quality of the display. This is especially true in high ambient light conditions where the light reflected by the display front surface can be many times greater than the display's black luminance when measured in a dark room.

Another, potential metric is Michelson Contrast, sometimes referred to as modulation. In this metric, we once again measure the luminance of a bright or white area, an area having a luminance L_h, and a dim area, an area having a luminance L_d. However, instead of computing a simple ratio of these values, a value is computed by calculating the ratio of the difference between these values and the sum of these values as shown in Eq. 2.2. The resulting value is in the range of 0–1.0. In this metric, displays with black levels that are extremely dark result in a contrast value near 1.0 and changes in this metric are more likely to correspond to our visual impression of a display. This metric is sometimes multiplied by 100 and expressed in percent contrast.

$$M = \frac{L_h - L_d}{L_h + L_d} \tag{2.2}$$

2.6 Defining Chromaticity

Although one can, and in some cases must, rely upon the addition of energy in the spectral domain to truly understand the utility of lighting and display systems, often times these calculations can be simplified significantly. The fact that human photopic vision relies predominantly on three sensors (i.e., cones), which is then processed by the human retina to provide three unique signals to the human brain, implies that the spectral information available from a light or display can be reduced to combinations of three numbers. These three numbers provide a simpler representation which is easier to interpret.

2.6.1 Tri-stimulus Values

As spectral data is transformed to the more simplified color space, it is important that properties, such as additivity, of the light be maintained. Additivity in color science is expressed by Grassmann's Laws; three empirical laws which describe the color-matching properties of additive mixtures of color stimuli. These laws, as summarized by Hunt, are as follows:

- To specify a color match, three independent variables are necessary and sufficient.
- For an additive mixture of color stimuli, only their tristimulus values are relevant, not their spectral compositions.
- In additive mixtures of color stimuli, if one or more components of the mixture are gradually changed, the resulting tristimulus values also change gradually.

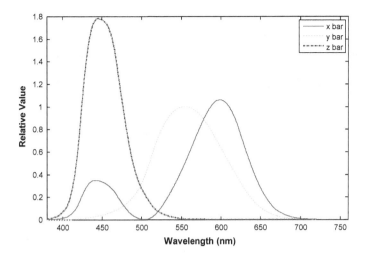

Fig. 2.7 CIE color matching functions useful in computing CIE Tristimulus Values X, Y and Z [4]

Although many transforms from spectral energy to this simplified color space could be applied, the most common transform is through application of the CIE 1931 Standard Colorimetric Observer functions to form CIE tristimulus values, typically denoted as X, Y, and Z. Tristimulus values are calculated by multiplying at each wavelength (i.e., convolving) the spectral energy of a source with each of three color-matching functions; $\bar{x}$, $\bar{y}$, and $\bar{z}$, shown in Fig. 2.7. Note that the shape of $\bar{y}$ is equivalent to the V(λ) function.

Calculation of the tristimulus values X, Y, and Z is performed using Eqs. 2.3 through 2.6:

$$X = K \sum_{\lambda=380}^{760} (S(\lambda)\bar{x}(\lambda)) \qquad (2.3)$$

$$Y = K \sum_{\lambda=380}^{760} (S(\lambda)\bar{y}(\lambda)) \qquad (2.4)$$

$$Z = K \sum_{\lambda=380}^{760} (S(\lambda)\bar{z}(\lambda)) \qquad (2.5)$$

where

$$K = \frac{100}{\sum_{\lambda=380}^{760} (S_w(\lambda)\bar{y}(\lambda))'} \qquad (2.6)$$

and where $S_w(\lambda)$ represents the spectral energy of the adapting white light source (e.g., the spectral energy of white on an emissive display or the measurement of reflectance from a perfectly white diffusor in a reflective environment) within the viewing environment and $S(\lambda)$ represents the spectral energy of the object being specified (i.e., a patch presented on an emissive display or an object in a reflective environment).

Although tristimulus values are useful as color remains additive within this color space, the values obtained are not especially meaningful. Larger values of Y (i.e., values near 100) generally imply the presence of relatively bright colors with respect to perfect white within the viewing environment and colors having near equal X, Y and Z values are generally neutral in color. Otherwise, it is generally difficult to interpret the values within this color space. Fortunately, these values can be transformed to more meaningful color spaces through simple transformations.

2.6.2 Chromaticity Coordinates and Diagram

One of the most frequently applied color spaces is the 1931 chromaticity space. This representation generally provides a transformation to a color space where colors can be understood by the numbers provided in this space.

The calculation of chromaticity coordinates (denoted by lower case x, y and z) are thus calculated from Tri-Stimulus values (denoted by upper case X, Y and Z) using the following transformations:

$$x = \frac{X}{X + Y + Z} \tag{2.7}$$

$$y = \frac{Y}{X + Y + Z} \tag{2.8}$$

$$z = \frac{Z}{X + Y + Z} \tag{2.9}$$

With this calculation, the sum of x, y and z will always equal 1. Therefore, only two (typically x and y) of the three values provide unique information. Color can then be represented in a two-dimensional CIE chromaticity diagram as was shown earlier in Fig. 1.5. This diagram typically contains the horse-shoe-shaped spectrum locus, representing the boundary of all possible colors. Colors near this spectral locus are generally highly saturated while colors distant from the spectral locus are muted. Values near the bottom left of this figure are blue, colors near the apex of the spectrum locus are generally green and colors near the bottom right are red. Referring back to Fig. 1.5, this figure also contained chromaticity coordinates for three example color display pixels. The colors enclosed by the triangle defined by the three coordinate pairs for the three colors of display pixels can be formed from different combinations of the three primary colors. Note that we have now lost all representation of luminance or brightness from this color space. Therefore colors which are differentiated by luminance (i.e., orange versus brown) cannot be differentiated within this diagram. Instead it becomes necessary to include a metric of relative lightness or brightness to fully describe colors.

The loss of luminance information also implies that we have lost the ability to directly add colors within this space. Instead, it now becomes necessary to include a luminance representation to add colors within this color space. For example, suppose we have two colors, each having chromaticity coordinates $(x_1, y_1; x_2, y_2)$ and luminance values m_1 and m_2. It can be shown that the resulting chromaticity coordinates x and y from adding the two colors can be calculated from the equations:

$$x = \frac{\frac{m_1 x_1}{y_1} + \frac{m_2 x_2}{y_2}}{\frac{m_1}{y_1} + \frac{m_2}{y_2}} \tag{2.10}$$

$$y = \frac{\frac{m_1 y_1}{y_1} + \frac{m_2 y_2}{y_2}}{\frac{m_1}{y_1} + \frac{m_2}{y_2}} \tag{2.11}$$

The geometric interpretation of this is that the new point color 3 (C_3) representing the mixture is on the line joining color 1 (C_1) and color (C_2) in the ratio calculated from:

$$\frac{C_1 C_3}{C_2 C_3} = \frac{\frac{m_2}{y_2}}{\frac{m_1}{y_1}}. \tag{2.12}$$

That is the new color lies at the center of gravity of weights calculated from the ratio of m_1 to y_1 for the first color (C_1) and the ratio of m_2 to y_2 for the second color (C_2). Therefore, Hunt refers to the result as the Center of Gravity Law of Color Mixture [10].

Finally, when observing the chromaticity diagram shown in Fig. 1.5, it has been shown that while relatively small differences in blue within this diagram are clearly visible, much larger changes in green must occur to have the same visual impact. It is therefore useful to develop methods to transform this color space to color spaces where the perceived differences in color are more uniform.

2.7 Uniform Color Spaces

The 1931 CIE Chromaticity space provides two highly desirable attributes. First, it permits us to describe any color with only three numbers; the luminance of the color, and the two chromaticity coordinates. Thus we can take any complex input spectra, including the one shown in Fig. 2.4 and transform it from the amount of energy at every wavelength to x, y and a luminance value. From these three values we can, relatively accurately, understand the perception of that color. Secondly, this color space provides us the ability to add colors, by simply adding luminance or computing the weighted average (i.e., center of gravity) of the x and y coordinates.

2.7 Uniform Color Spaces

Further, the three numbers that are created are somewhat interpretable. For instance colors with low values of x and y will appear blue in color, colors with high values for x but low values for y will appear red in color, and colors with intermediate values of x but high values for y will appear green in color. Finally, colors near equal energy white (0.33, 0.33) will generally appear white in color. Each of these attributes lend significant value to the CIE 1931 chromaticity space.

Unfortunately, the CIE 1931 chromaticity space, as alluded to earlier, has a significant drawback. Specifically, it is not perceptually uniform. This deficiency can be troubling, especially when one wishes to understand the perceived impact in color changes. That is, if you wanted to ask the question: "Could a person see a difference in color if the distance in x, y space changed by 0.04 units?" The answer is, it depends on where the original color resides within the 1931 chromaticity space. If it is in the blue area, then the answer is certainly. However, if the original color is in the green area, then the answer is certainly not. Thus efforts were undertaken to derive alternate chromaticity spaces which overcome this deficiency.

2.7.1 CIE 1976 Uniform Chromaticity Diagrams

It is possible to transform the chromaticity color space to a perceptually more uniform color space. One such color space was adopted and is known as the CIE 1976 uniform chromaticity scale diagram or the CIE 1976 UCS diagram, often referred to as the u'v' chromaticity diagram. This chromaticity space can be obtained by transforming among the previously discussed color spaces using the equations:

$$u' = \frac{4X}{X + 15Y + 3Z} = \frac{4x}{-2x + 12y + 3} \tag{2.13}$$

$$v' = \frac{9Y}{X + 15Y + 3Z} = \frac{9y}{-2x + 12y + 3} \tag{2.14}$$

As with the CIE 1931 chromaticity color space, color can be added through application of the Center of Gravity Law of Color Mixture.

To further define color, the CIE further defined metrics for hue angle and saturation. Hue is an attribute of an object indicating the degree to which it appears similar to one, or as a proportion of two, of the colors red, yellow, green, and blue. Hue angle is a metric of hue (denoted h_{uv}) which represents the angle of the vector formed between the coordinates of white and the target color. Hue angle is calculated as:

$$h_{uv} = \tan^{-1}\left(\frac{v' - v'_n}{u' - u'_n}\right) \tag{2.15}$$

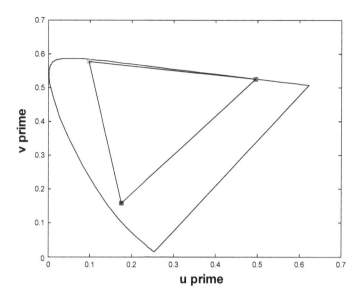

Fig. 2.8 CIE 1976 Uniform Chromaticity Scale Diagram containing the same example display primaries as shown in Fig. 1.5

where v′ and u′ are the chromaticity coordinates for the color and v_n' and u_n' are the uniform chromaticity coordinates for the white point or neutral point within the scene.

Saturation represents the purity of the color with colors near white having low saturation and purer colors being higher in saturation. Therefore saturation (denoted s_{uv}) is calculated by determining the distance of the color from white or neutral using the equation:

$$s_{uv} = 13\sqrt{\left(u' - u_n'\right)^2 + \left(v' - v_n'\right)^2} \qquad (2.16)$$

where u_n' and v_n' represent the UCS coordinates for white.

With these metrics, a set of coordinates in a two-dimensional space exist which overcome the primary deficiency of the 1931 chromaticity space. Specifically, the chromaticity coordinates are perceptually uniform. Figure 2.8 shows the same display primaries as shown in the chromaticity diagram of Fig. 1.5. However, in Fig. 2.8, these display primaries have been rendered into the 1976 UCS diagram for illustration purposes. As shown in this figure, the u-shaped chromaticity space has been rotated and the curved portion of the u has been condensed to reduce the separation of the green chromaticity coordinates. However, the three display primaries still form a triangle and the colors within this triangle can be formed through different combinations of luminance from the three primaries.

Applying this chromaticity space, we have a color space that is uniform in two dimensions. However, remembering our earlier discussion of luminance, you may

2.7 Uniform Color Spaces

recall that our eye is a logarithmic sensor with respect to luminance. Therefore, we still lack a three-dimensional uniform color space.

2.7.2 1976 CIE L*u*v* Uniform Color Space

To overcome this problem, uniform color spaces, such as the 1976 CIE L*u*v* uniform color space was formulated in an attempt to provide a three-dimensional uniform color space. This scale provides a color space that is referenced to the white point of the image or display. To define this space, we will begin by defining lightness, which corresponds to the relative brightness of a color with respect to the white point.

The lightness scale attempts to scale luminance such that an equal increase in lightness is equivalent to a perceptually equal change in perceived lightness where 100 corresponds to a color that appears equal to the white point. This value is calculated as follows:

$$L^* = 116 \sqrt[3]{\frac{Y}{Y_n}} - 16 \; for \; \frac{Y}{Y_n} > 0.008856 \quad (2.17)$$

and

$$L^* = 903.3 \frac{Y}{Y_n} \; otherwise \quad (2.18)$$

In this color space, perceived lightness is computed as the cubed root of the ratio of luminance to the luminance of white. Although this is not a logarithmic function, it provides a similar transform, where L* increases rapidly for changes in low luminance values and less rapidly for similar changes in high luminance values.

While u'v' represents a uniform color space and L* scales luminance such that equal changes in L* are equal in perceived magnitude, these two sets of values are not on the same scale. That is a change in L* of 1 does not represent the same perceptual differences as a change in u'v' of 1. Therefore the CIE additionally defined a uniform color space, where L* is calculated to indicate a change in luminance and u* and v* are calculated as follows:

$$u^* = 13L^* \left(u' - u'_n \right) \quad (2.19)$$

and

$$v^* = 13L^* \left(v' - v'_n \right) \quad (2.20)$$

With this, a change in L*, u* or v* represents about an equal change in color. We can then compute a simple Euclidean distance in this three-dimensional space which represents a change in color or a color difference value. This metric is defined as follows:

$$\Delta E^*_{uv} = \sqrt{\Delta L^{*2} + \Delta u^{*2} + \Delta v^{*2}} \qquad (2.21)$$

2.7.3 1976 L*a*b* Uniform Color Space

Similar to the L*u*v* color space, the L*a*b* color space permits calculation of color in a space which is intended to be perceptually uniform. Once again, L* is calculated according to Eqs. 2.17 and 2.18. The a* and b* values in this color space are calculated as:

$$a^* = 500\left(\sqrt[3]{\frac{X}{X_n}} - \sqrt[3]{\frac{Y}{Y_n}}\right) \qquad (2.22)$$

$$b^* = 200\left(\sqrt[3]{\frac{Y}{Y_n}} - \sqrt[3]{\frac{Z}{Z_n}}\right) \qquad (2.23)$$

where Xn, Yn, and Zn are the X, Y, Z values for the appropriately chosen reference white, when the ratios of X/Xn, Y/Yn and Z/Zn is greater than 0.008856. If any of these ratios is less than 0.008856, then the ratio is calculated using 7.787F + 16/116, where F is the ratio.

As with the L*u*v* space, a color difference can be computed as a Euclidean distance in a three-dimensional space, using an equation similar to Eq. 2.21, where u* and v* are replaced with a* and b*.

Each of the distance metrics can be applied by simply calculating the difference in values computed for two colors to assess the degree of difference that is present. It is generally accepted that a value of 1 is near a "just noticeable difference", that is 50% of observers should be able to detect this difference under constrained viewing conditions. A value of 3 is required before the color will be apparent to a casual observer.

2.8 General Color Rendering Index

Similar metrics can be constructed to answer more complex questions. For example, when assessing a man-made light source, one can ask how well the color output by the light source mimics or represents natural light. However, we do not see the color of light as it passes through space but as it radiates from its emissive source or is reflected from objects in our environment. Therefore, we might modify this question to ask: "Do objects illuminated with the man-made light source appear equivalent

2.8 General Color Rendering Index

in color to the same objects illuminated by a desired natural light?" One metric for accomplishing this is the General Color Rendering Index (R_a), standardized by the CIE in 1965. In this metric, a value R_a is calculated which varies between 0 and 100, with 100 indicting that the manmade light performs equivalent to the reference light. This value is computed from the following equation:

$$R_a = 100 - \frac{4.6}{k} \sum_{i=1}^{k} d_i \quad (2.24)$$

where d_i is a color difference metric designed to compare the apparent color of a series of k standard objects when illuminated by the light source being evaluated as compared to the apparent color of the same standard objects when illuminated by a desired natural light source. Note that in this metric, if d_i is equal to zero for all standard objects, the value of the metric is 100. However, as the difference in apparent color between the standard objects lit by the light being evaluated and the standard objects lit by the natural light increases, the resulting value decreases. Thus the maximum color rendering index value is 100 and poorer quality light sources will provide a lower color rendering index value. The difference metric applied in the color rendering index calculation is shown in the following equation:

$$d_i = 800\sqrt{\left(u'_i - u'_{ni}\right)^2 + \left(v'_i - v'_{ni}\right)^2} \quad (2.25)$$

where this computation is performed for k standard patches (the standard defines 8 standard color patches and 6 supplemental patches) and n represents the values produced by the reference (natural) light source [11].

2.9 Defining Human Needs for Color

Thus far we have briefly discussed the mechanisms in the human visual system which influences our perception of color and some ways to begin to quantify our perception of color. We have not discussed how we use or why we care about color. Early color display research illustrated that color has many functions. It aids visual search, helping us to more rapidly locate objects with saturated colors in our environment [3]. Additionally, it aids our ability to segment and to recognize objects more rapidly [13]. The fact that recognition is aided by color is important because it supports the fact that we remember the colors of objects and use this information to aid recognition. Studies of human performance which illustrate the utility of color are important, as they help us to understand that color is useful to our everyday existence.

> The human visual system defines color with respect to white, regardless of whatever "white" actually is at the current moment.

So if color aids our ability to recognize objects then we must remember the color of objects, associating these objects with certain colors. Of course, this makes intuitive sense to us given our everyday experience. After all, grass is green, skies are blue, and clouds are white. However, as we will see in upcoming chapters, the physical color of daylight changes dramatically throughout a single day. This implies that the light reflected from that cloud changes as well and that the physical color of the cloud must, therefore, change dramatically throughout the day. However, it remains this color that we call white rather than being yellow in the morning and blue in the afternoon as physical measurements of the light reflected from the cloud might lead a scientist to believe should be the case. The reason the cloud appears constantly white is adaptation. Our visual system analyzes the scene to determine objects that should appear white and then determines color with respect to that white color. This is captured in the use of the neutral or white reference in the calculation of the tristimulus values as well as most of the uniform color space metrics. The fact that all color in our visual world is referenced to this white color is a very important concept in color science. The fact that the perceived colors of objects in a scene is influenced by the white color in the scene is a clear indication that color, like beauty, is truly in the eye of the beholder and does not exist as a physical entity. It is only by modeling basic attributes of human color vision that we acquire the ability to measure, quantify and mathematically describe color.

> Generally individuals prefer images or displays which produce images having more highly saturated colors and higher levels of contrast.

So if white is not a color with a single, given set of chromaticity coordinates, but one of several possibilities within an area of the chromaticity diagram, is this also true of the green of our leaves? Of course, this is true as well for green and most other saturated colors. These colors are referenced to white as the color of white changes throughout the day in our natural, outdoor environment. However, it is maybe even more true of the green associated with tree leaves as there is a lot of natural variability associated with green tree leaves as different trees have different colors of leaves, the reflectance of leaves on a single tree vary throughout the year and the reflectance of the leaves vary with the health of the tree. Each of these sources of variability add to the range of possible colors that we associate with tree leaves. Therefore, studies have shown that individuals prefer images having more highly saturated colors than less saturated colors, as long as the hue associated with the color does not vary over a large range and spatial artifacts are not introduced [28]. That is, as long as we do not change the color of leaves such that they begin to appear abnormally

2.9 Defining Human Needs for Color

blue or yellow, generally increasing the saturation of the tree leaves is appealing. This increase in saturation improves the perceived contrast between objects within images and increased contrast generally improves perceived image quality [18].

> Saturation increases are not desirable for near-neutral and flesh colors.

While increasing the saturation of most colors is desirable from an image quality point of view, this is not true of all colors. Particularly near neutral colors and colors associated with skin (i.e., flesh colors) must be rendered without significant changes in either hue or saturation. Colors near white of course help to reinforce the white point and because our visual system perceives color as differences, it is relatively sensitive to even small changes in colors near neutral. Flesh is important as its color helps us understand the emotional level and health of our friends and family. Thus we are highly sensitive to changes in the color of flesh as we associate even subtle increases in saturation with increased blood flow and flushing associated with fever or a stress reaction.

From a display perspective, we then value display systems which provide precise rendering of neutral (near white) colors as well as the ability to create highly saturated colors. We then prefer rendering algorithms which manipulate image data to produce highly saturated colors on these displays by boosting the chroma of colors by as much as 20% while rendering white, flesh, and other neutral colors reliably [20].

2.10 Summary and Questions to Consider

Approaching the end of this chapter, we see that generally the human visual system adapts to our natural surroundings. As we will discuss further in the next chapter the color, as well as the intensity, of light varies significantly throughout the day and yet because of the adaptation of our visual system, we are not aware of these changes. Light not only affects our perception of the world but the way our body reacts to the world. Each of these effects are quite complex. As a result, the color science community has constructed several, increasingly complex models to permit us to perform physical measurements and calculations to provide insight into how we will perceive an environment. We reviewed some of the basic models in this area, although more complex models exist that I have intentionally avoided discussing these metrics for the moment. Never the less, these relatively simple mathematical representations provide useful metrics that we will employ throughout this text. These metrics provide methods to compute values which tell us something about the perceived color or the difference between two colors where these colors exist within a three-dimensional space. They also provide a basis that can provide insight into higher level human percepts, such as preferred color and image quality. With this review, here are a few questions for you to ponder.

1. If most of our cones are located in our fovea, how is it that color significantly improves our ability to perform visual search across a large field of view?
2. Given that we only see high resolution information in a small area of our field of view and that we are generally not aware of this restriction, what does this imply about our true knowledge of the world or a displayed image or a displayed video?
3. Given that it takes our visual system time to adapt to darkening conditions and less time to adapt to brighter environments, what are the implications for image processing systems?
4. We have provided a standard set of color metrics but we must also recognize that we all vary somewhat in our visual abilities, for example some individuals exhibit varying degrees of color blindness. How accurate are these metrics both within and between people?
5. It is recognized that aging and exposure to light can cause yellowing of our cornea, degrades the sensors, especially the blue sensitive sensors in our eyes, and can lead to losses of neural function. How might these effects alter the accuracy of these metrics for predicting the perception of older individuals?

References

1. Association National Electrical Manufacturers (2011) Digital imaging and communications in medicine (DICOM) part 14 : grayscale standard display function, Rosslyn, VA. Retrieved from http://www.ncbi.nlm.nih.gov/pubmed/2188123
2. Berson DM, Dunn FA, Takao M (2002) Phototransduction by retinal ganglion cells that set the circadian clock. Science (New York, N.Y.) 295(5557):1070–1073. http://doi.org/10.1126/science.1067262
3. Christ RE (1975) Review and analysis of color coding research for visual displays. Hum Factors 7(6):542–570
4. CIE (1932) Commission internationale de l'Eclairage proceedings. Cambridge University Press, Cambridge
5. Coren S, Porac C, Ward LM (1984) Sensation and perception (Second). Academic Press Inc, Orlando, FL
6. Dawson D, Campbell SS (1991) Timed exposure to bright light improves sleep and alertness during simulated night shifts. Sleep 14(6):511–516
7. Foster RG, Provencio I, Hudson D, Fiske S, De Grip W, Menaker M (1991) Circadian photoreception in the retinally degenerate mouse (rd/rd). J Comp Physiol A 169(1):39–50. https://doi.org/10.1007/BF00198171
8. Gouras P (1991) Precortical physiology of colour vision. In: Cronly-Dillon J (ed) The pereption of colour: vision and visual dysfunction, vol 6. CRC Press Inc, Boca Raton, FL, pp 179–197
9. Hennessy RT, Iida T, Shiina K, Leibowitz HW (1976) The effect of pupil size on accommodation. Vis Res 16(6):587–589. http://doi.org/https://doi.org/10.1016/0042-6989(76)90004-3
10. Hunt RWG (1995) The reproduction of colour, 5th edn. Fountain Press, Kingston-upon-Thames England
11. International Commission on Illumination (CIE) (1974) Method of measuring and specifying color rendering properties of light sources: CIE 13.3:1974
12. International Commission on Illumination (CIE) (2010) Recommended system for mesopic photometry based on visual performance: CIE 191:2010

References

13. Janssen TJWM (1999) Computational image quality (Unpublished Doctoral Dissertation). Eindhoven University of Technology
14. Kardon R, Anderson SC, Damarjian TG, Grace EM, Stone E, Kawasaki A (2009) Chromatic pupil responses: preferential activation of the melanopsin-mediated versus outer photoreceptor-mediated pupil light reflex. Ophthalmology 116(8):1564–1573. https://doi.org/10.1016/J.OPHTHA.2009.02.007
15. Lockley SW, Evans EE, Scheer FAJL, Brainard GC, Czeisler C, Aeschbach D (2006) Short-wavelength sensitivity for the direct effects of light on alertness, vigilance and the waking electroencephalogram in humans. Sleep 29:161–168
16. Loschky L, McConkie G, Yang J, Miller M (2005) The limits of visual resolution in natural scene viewing. Vis Cogn 12(6):1057–1092. https://doi.org/10.1080/13506280444000652
17. Lucas RJ, Peirson SN, Berson DM, Brown TM, Cooper HM, Czeisler CA, Brainard GC (2014) Measuring and using light in the melanopsin age. Trends Neurosci 37(1):1–9
18. Miller ME (1993) Effects of field of view, MTF shape, and noise upon the perception of image quality and motion (Unpublished Doctoral Dissertation). Virginia Tech
19. Mosaly PR, Mazur LM, Marks LB (2017) Quantification of baseline pupillary response and task-evoked pupillary response during constant and incremental task load. Ergonomics 60(10):1369–1375. https://doi.org/10.1080/00140139.2017.1288930
20. Murdoch MJ (2013) Human-centered display design balancing technology and perception (Unpublished Doctoral Dissertation). Eindhoven University of Technology
21. Okamoto Y, Rea MS, Figueiro MG (2014) Temporal dynamics of EEG activity during short- and long-wavelength light exposures in the early morning. BMC Research Notes. http://doi.org/10.1186/1756-0500-7-113
22. Shorter PD (2015) Flashing light-evoked pupil responses in subjects with glaucoma or traumatic brain injury (Unpublished Doctoral Dissertation). The Ohio State University
23. Thapan K, Arendt J, Skene DJ (2001) An action spectrum for melatonin suppression: evidence for a novel non-rod, non-cone photoreceptor system in humans. J Physiol 535(1):261–267. https://doi.org/10.1111/j.1469-7793.2001.t01-1-00261.x
24. Tryon W (1975) Pupillometry: a survey of sources of variation. Pscyhophysiology 12(1):90–93
25. Vollmerhausen RH, Jacobs E (2004) The targeting task performance (TTP) metric a new model for predicting target acquisition performance. Fort Belvoir, VA
26. Winn B, Whitaker D, Elliott DB, Phillips NJ (1994) Factors affecting light-adapted pupil size in normal human subjects. Invest Ophthalmol Vis Sci 35(3):1132–1137. Retrieved from http://www.ncbi.nlm.nih.gov/pubmed/8125724
27. Wyszecki G, Stiles WS (1982) Color science: concepts and methods, quantitative data and formulae, 2nd edn. Wiley, New York, NY
28. Yendrikhovskij SN (1998) Color reproduction and the naturalness constraint (Unpublished Doctoral Dissertation). Eindhoven University of Technology

Chapter 3
Scenes and Lighting

3.1 Measurement of Light

As we think about the ways that light can enter our eyes, there are two general paths. First, we can look at an object which is emitting light, such as a light bulb or an electronic display. Second, we can look at an object which reflects light, like a piece of paper or the leaf on a tree. In either case, the light source or the reflector is giving off light which we can measure.

3.1.1 Luminance Measurement

Luminance is used to measure the light which is being given off by a surface. To make this measurement, we might employ a meter such as the one shown in Fig. 3.1 to measure the light leaving the surface of a display or a reflective surface before the light enters our eye. Under these conditions we might physically measure the amount of energy (measured in joules) given off by a point on the surface of the light emitter or the reflector for a period of time (measured in seconds). Therefore, this is measured in units of J/s or Watts. Note that in this case, we are measuring the energy given off by the surface in a particular direction and we typically measure the light collected over an angle, projected onto sphere, which is expressed in units of steradian (sr). For example, a typical radiometer might measure the amount of energy given off by the surface over an angle of 1° with the assumption that this energy is projected onto a 1 m^2 area of a sphere with radius of 1 m around the point source. In this measurement, we are measuring optical power or radiance given off by the surface.

Once this measurement is made, each wavelength of light can be convolved with the appropriate eye sensitivity function and summed across all wavelengths to produce a luminance value. Luminance is specified in units of lm/sr/m^2,

© Springer Nature Switzerland AG 2019
M. E. Miller, *Color in Electronic Display Systems*, Series in Display Science and Technology, https://doi.org/10.1007/978-3-030-02834-3_3

Fig. 3.1 Illustration of a typical spectral radiometer showing an output spectrum from flat source. Picture used with permission from JADAK, a business unit of Novanta

typically indicated in terms of candela per square meter, abbreviated cd/m^2 where the unit candela is defined as 1 lm per steradian [15]. Regardless of which path the light travels before entering our eye, our perception of brightness of an object within an environment will be correlated with luminance measured in cd/m^2, provided through this approach.

3.1.2 Illuminance Measurement

At times, we do not want to know how much light is given off by a surface. Instead, we want to know how much light is actually hitting a surface. For example, we might want to know how much energy from our overhead lights is hitting the book or display you are using to read this text. In this case, we will likely use an illuminance meter, such as shown in Fig. 3.2. As shown in this figure, a typical illuminance meter will collect light from a half sphere collecting all light that will hit the surface, regardless of the direction of the light. Again, the light gathered by the device is weighted by an eye sensitivity function. In this circumstance, the amount of light is measured in terms of lumens per square meter, often referred to as Lux and abbreviated as lx.

3.1 Measurement of Light

Fig. 3.2 Illustration of a typical illuminance meter arranged to perform an illuminance measurement of light impacting a surface

Luminance is a measurement of the visible light given off by a surface. Illuminance is a measurement of the visible light which impacts a surface.

At times, we may also wish to understand the relationship between the illuminance of a surface and the luminance of that same surface. However, to understand this relationship, we have to have knowledge of the direction of the light as it contacts the surface, the reflectance of the surface as it is illuminated by the light source from the given angle and the angle at which the surface is viewed. Therefore, exact specification of this relationship is complex unless the reflectance of the surface, the lighting environment, and the viewing environment are specified precisely or simplifying assumptions are made.

3.1.3 Power, Efficiency, and Efficacy

As we prepare to talk about measuring light from manmade light sources and displays we also need to clearly define three terms useful in understanding the utility of the energy that the light sources or displays produce. Generally, we are concerned about the efficiency with which we can convert electrical energy to light. To specify this efficiency, we can begin by measuring the electric power consumed by a device (P_e) using traditional power measurement devices. We can then measure the optical power (P_o) output by the device. This optical power is usually measured with a radiometer, often equipped with an integrating sphere. The integrating sphere collects the light emitted by the light source, regardless of the direction of the light and permits measurement of this light using a spectral radiometer, such as the device shown in Fig. 3.1. When summed across all wavelengths we have the total optical power output by the light source. Electrical efficiency is then the ratio of the optical power created by the light source to the electrical power consumed by the device. That is efficiency (E) is calculated as shown in Eq. 3.1.

$$E = \frac{P_o}{P_e} \qquad (3.1)$$

However, it is not power which is important to the human eye, but luminance. Therefore, what we often wish to know is how efficient is the device at producing luminance, to which the eye is sensitive. That is we want to understand the ratio of the luminance (L) output by the device to the electrical power input to the device (P_e). This term is referred to as efficacy (E_a) of the source and is calculated as shown in Eq. 3.2.

$$E_a = \frac{L}{P_e} \qquad (3.2)$$

Note that if the eye was equally sensitive to all power, E and E_a would be equal. However, the eye is only sensitive to a certain frequency band of optical power and is much more sensitive to optical power in the yellow portion of the visible spectrum than for shorter or longer wavelengths. Therefore, E and E_a are often quite different from one another, with E_a providing a better estimate of the utility of the light to the user.

3.2 Daylight

Natural light is provided by the sun throughout the day. Importantly the light the sun provides at the surface of the earth changes in direction, intensity, and color throughout the day and each day of the year. As the sun rises in the eastern sky each morning, it appears near the horizon. At this location with respect to our location on the earth's surface, the light rays from the sun enter the earth's atmosphere and travel a substantial distance through the atmosphere. As such, water vapor and particles in the atmosphere absorb a substantial portion of the short-wavelength, high energy portion of the sun's energy before it reaches us. Therefore, the intensity of the light is reduced and the light which reaches us has most of the blue and other short wavelength energy filtered out of it, leaving a yellow-ish white light. As the earth rotates and the

3.2 Daylight

sun enters the atmosphere from overhead around midday, the energy from the sun travels a substantially shorter distance through the atmosphere. Thus, the intensity of the sun increases and less of the short wavelength light is filtered out of sunlight, permitting it to provide a blueish-white light. Finally, as the sun drops towards the horizon in the afternoon, the distance its light rays travel through the atmosphere lengthens; which reduces the intensity and shifts the color of light towards yellow again. As a result, we can expect significant changes in light intensity and color of illumination from the sun throughout each day. Clouds also have an effect on the quality of the light at the earth's surface, typically reducing the intensity of light but often reducing the long wavelength energy, creating bluer light.

The intensity of light of course varies over a large range each day. Daylight can provide illuminance values on the order of 100,000 lx at midday. This intensity decreases as the sun drops below the horizon such that a starlit night might have intensities as low as 0.002 lx. Therefore, the intensity of this illumination changes over a very large range.

> Both the intensity and color of daylight changes dramatically throughout a typical day.

As we talk about the color of light, one way to discuss this color is through the use of correlated color temperature. However, to understand this term, we first need to understand the Planckian locus.

Physicists have defined a black body as an object that absorbs all radiation falling on it, at all wavelengths. When a blackbody is heated, it produces photons with a characteristic frequency distribution, which depends on its temperature. At lower temperatures, these photons are low in energy and therefore have long wavelengths, resulting in red light. As the temperature is increased, shorter wavelength photons are produced and the color of the light emission shifts through white towards a blueish-white. The color of the resulting light emission can plotted on the Planckian locus within a uniform chromaticity diagram as shown in Fig. 3.3. Each point along this curve then corresponds to a different temperature of the blackbody radiator [15].

Correlated color temperature is then determined by drawing a line perpendicular to the Planckian locus from the chromaticity coordinate of the light source to the color temperature of the light of interest. The resulting point of intersection of the line with the Planckian locus is referred to as the correlated color temperature. This metric provides an imprecise metric of the color of illumination as it does not specify the distance or length of the line and thus does not completely specify the chromaticity of the light source. However, as long as the point of interest is near the Planckian locus, it provides a useful indication of the perceived color of the light source.

Because the sun is not exactly a blackbody radiator, a curve which separate, but similar to the Planckian locus can be constructed to show the colors of standard daylight as shown in Fig. 3.3. Note that this curve does not represent all possible colors of daylight but represents a standard set of colors of daylight measured at the earth's sur-

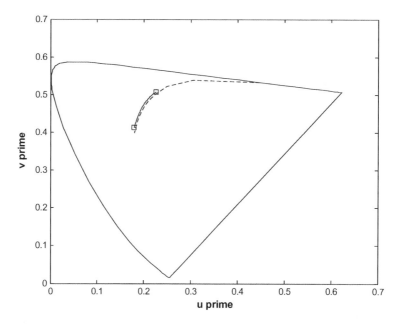

Fig. 3.3 A 1976 UCS diagram illustrating the location of the Plankian locus (shown as a dashed line) and the standard daylight curve for color temperatures between 3850 and 25,000 K (shown as solid line with squares indicating end points)

face (International Commission on Illumination (CIE) [6]. Researchers have shown that the correlated color temperature of daylight varies between approximately 3700 and 35,000 K throughout the day [4], with the higher color temperature occurring near midday with clouds in the sky. However, standard daylight curves have been defined between 3850 and 25,000 K as shown in Fig. 3.3. Note that the daylight curve lies near and approximately parallel to the Planckian locus.

In addition to specifying the standard daylight curve, standard daylight spectra have also been adopted. Figure 3.4 shows the standard spectra for three standard daylight curves. As one can see, these curves are relatively continuous. That is, they each include some energy at every visible wavelength and the amount of energy changes gradually as wavelength changes within each curve. However, the 11,000 K curve, which represents midday sunlight, contains predominantly short wavelength (blue) energy while the 3850 K curve, which represents morning and evening sun, contains predominantly long wavelength (yellowish-red) energy. The 6500 K daylight curve, contains approximately equal energy across the visible spectrum while maintaining the general shape of the daylight curve.

Recalling our discussion of human vision, the ipRGCs are sensitive primarily to high intensity, short wavelength, energy. As such, these sensors are likely to be highly sensitive to the very high intensity, high color temperature midday sun and not as sensitive to the lower intensity, low color temperature sun in the morning or evening.

3.2 Daylight

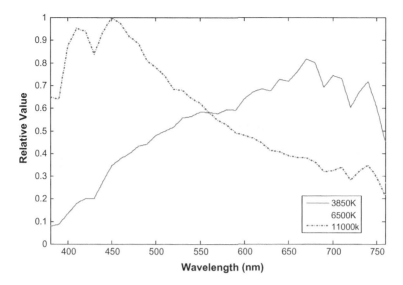

Fig. 3.4 Standard daylight spectra corresponding to 3870, 6500 and 11,000 K. Note the preponderance of short wavelength (blue) light in the 11,000 K curve and the preponderance of long wavelength (yellowish-red) light in the 3850 K curve. Curves plotted from tabular data provided in Wyszecki and Stiles [15]

As it is believed that the signals from these cells reduce the production of melatonin in the pituitary gland, reducing our stimulus to sleep, the intensity and color of natural daylight not only affects our perception of the environment, but likely plays a significant role in regulating our sleep patterns and our level of alertness. Thus, while we may not be aware of the changing color and intensity of natural light throughout the day, these changes likely have other significant effects on our performance, sleep, and long term health.

Within the display industry today, it is important to note that there is significant concern that the blue intensive white light emitted by popular displays may be producing changes in circadian rhythms. While the science behind this concern is still progressing, it is important to realize that the color of the light is one of the attributes of daylight which changes with time of day. It is important to realize that the intensity of the light from midday to evening can change by several orders of magnitude and the luminance output by current day displays falls far short of the luminance reflected by our natural surroundings at midday.

3.3 Characteristics of Artificial Lighting

Artificial or manmade light permits us to function in enclosed spaces and during non-daylight hours by providing an alternate source of illumination. Although there are many sources available today, much of internal home and office lighting has been provided by either the traditional light bulb, called incandescent lighting, or fluorescent lighting. In recent years, Light-Emitting Diodes (LEDs) have begun to replace most traditional manmade light sources. In this section, we will quickly review the characteristics of incandescent and fluorescent lighting, then take a more detailed look at LEDs for lighting. Interestingly, fluorescent and LEDs not only serve to provide illumination sources in manmade environments but also provide an illumination source in some displays, for example within liquid crystal displays. We will, therefore, discuss some characteristics of each of these technologies within each of these applications.

3.3.1 Incandescent

The traditional incandescent bulb contains a thin piece of material having a high electrical resistance, typically tungsten. This high resistance material is referred to as the filament. As electricity is passed through the filament much of the current is converted to heat because of the high resistance. Eventually the filament is heated to a high temperature and begins to radiate a portion of the energy in the form of photons. However, in this process, most of the energy is converted to and radiates from the bulb as heat. Therefore, often less than 20% of the energy input to the incandescent bulb is converted to light with the remaining converted to heat.

This device is relatively simple as its primary components are the filament and electrical connections to the power source. However, to prevent the tungsten from oxidizing, which would cause rapid degradation, this filament must be housed in a vacuum. The familiar light bulb is then used to maintain an enclosure for the vacuum while permitting the light from the filament to be emitted. An important characteristic of this light source is that light is emitted in virtually all directions.

> The output spectra of incandescent bulbs are fixed. Therefore, these bulbs output light with a fixed color output having a single correlated color temperature.

The filament in a traditional light bulb is heated to about 2700 K and behaves much as a blackbody radiator, emitting energy with a color temperature of about 2700 K. Figure 3.5 shows a typical incandescent spectra. The light produced by an incandescent bulb thus has a distinct yellow tint. Our eye can largely adapt to this color of light and therefore items within an environment can have nearly a full range of color. However, there is a decidedly lack of short wavelength light emitted from these bulbs. Therefore, objects which reflect predominantly short wavelength, blue light will have little energy to reflect when illuminated by an incandescent bulb. Blue objects, when lit with incandescent light, will appear dark and less vibrant than they would in environments lit with higher color temperature light. Studies have shown that under low color temperature lighting, individuals prefer furniture and decorations

3.3 Characteristics of Artificial Lighting

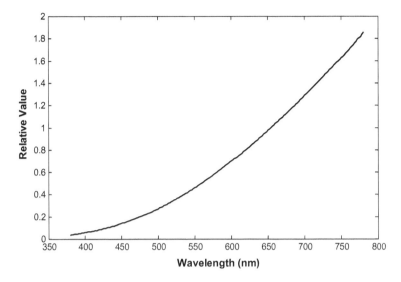

Fig. 3.5 Spectral emission from a typical incandescent bulb

containing substantial amounts of reds, yellows and oranges; likely because these colors will reflect more of the energy emitted by incandescent lighting and appear brighter and more vibrant than greens or blues. Thus in manmade environments, which are lit predominantly by incandescent lighting, most objects can be expected to exist which reflect the longer wavelength energy provided by incandescent lighting.

With any lamp, it is important to understand how well it represents natural lighting. Remembering the color rendering index from the last chapter, if we use daylight having a color temperature near the color temperature of the incandescent bulb as the reference light, we will see that the color rendering index of the incandescent bulb is quite high, often above 95, as the shape of emission is smooth and quite similar to the corresponding daylight spectra.

An important attribute of incandescent bulbs is that the lifetime of these bulbs is somewhat dependent upon the number of on-off cycles as the filament can fracture while undergoing the extreme temperature changes, which occur as the bulb is cycled on or off. Additionally, a short period of time is required to heat the filament before it emits light. Therefore, incandescent light sources are not cycled but are activated for use and not turned off until they are no longer needed.

3.3.2 Fluorescent

Fluorescent bulbs do not have a filament but are filled with a gas containing mercury. When this gas is exposed to a high voltage, the vapor is heated and begins to emit high energy photons. The inside of the bulb is coated with a phosphor coating. When these phosphors are bombarded with the high energy photons produced by the gas, the molecules within the phosphor reach a high energy state and then produce both heat and a lower-energy photon to obtain equilibrium. Typical fluorescent bulbs are coated with a few specifically-designed phosphors, each of which emit photons at a single frequency. Therefore, the bulbs emit visible light at only a few peaks within the visible spectrum. The color of the light and the bandwidth of each emission peak can be varied by coating the inside of the bulb with different phosphors. The power efficacy and cost of the bulb can be significantly affected by the selection of phosphors. Therefore, most bulbs contain a standard set of narrow-band phosphors.

Generally, fluorescent lighting is more energy efficient than incandescent. Typical fluorescent bulb efficacy is in the range of 40–50%. Therefore, they are often more than two times as efficient as incandescent bulbs at converting electrical energy to useful light. In western societies, where energy concerns have traditionally not been severe, these bulbs have found use predominantly in large area commercial locations but have not replaced incandescent lighting within homes. However, in cultures where energy is of greater concern, fluorescent lighting has, until recently, been used in most indoor environments.

Figure 3.6 shows a typical emission spectra for a common warm fluorescent bulb. The light from fluorescent bulbs is often decidedly "bluer" than the light produced by incandescent bulbs. However, warm fluorescent manufactured for in-home use can have a lower, often referred to as "warmer", color temperature. For example, the spectra shown in Fig. 3.6 were measured from a fluorescent bulb having a color temperature around 2800 K. Note, that the emission spectra appear decidedly different than the natural or incandescent spectra. Both daylight and tungsten spectra are relatively smooth having some energy at virtually every wavelength within the visible range. The emission from fluorescent bulbs is characteristically different containing significant energy at a small number of wavelengths, with much less energy at other wavelengths. Using daylight spectra with the same color temperature as the corresponding fluorescent lamp, we will see that the color rendering index for these lamps will often be lower than for incandescent bulbs because of the peaked response of these lamps. While they can be produced to provide color rendering index values greater than 80, the lack of a smooth emission band compromises the reproduction of some colors within the environment.

Historically, the discrete nature of emission from fluorescent bulbs was a concern. Thus significant research was conducted on "full spectrum" fluorescent bulbs to understand if these bulbs might have positive effects on vision, perception or human health. The industry, however, settled on the use of fluorescent bulbs with a peaked spectrum as the energy and cost savings from this approach was deemed more valuable than any potential benefit from "full spectrum" fluorescent bulbs.

> The emission spectra of a fluorescent bulb are fixed by the phosphors coated in the bulb. Therefore, the color of light output by a fluorescent bulb is constant once they are manufactured. However, the color of light can be influenced by phosphor selection.

3.3 Characteristics of Artificial Lighting

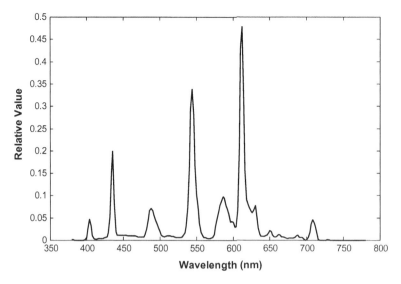

Fig. 3.6 Spectral emission for a typical fluorescent bulb

One other note is that fluorescent bulbs require a noticeable time to be activated. This is particularly true in low temperatures as the vapor in the bulb must be heated before light emission occurs. Therefore, like incandescent, it is not practical to rapidly cycle this source from on to off.

Fluorescent bulbs have been used not only as a source of light for room lighting but can be used as a light source in liquid crystal displays. The characteristics of these bulbs in this application are similar to general illumination. They require a very high voltage to function and are not pulsed once they are turned on. Additionally, these bulbs are often large, extending across the full width or height of a display. Multiple fluorescent bulbs are then placed behind a display to provide even illumination across the display.

3.3.3 LED

Inorganic Light Emitting Diodes (iLEDs) are currently the LEDs that many of us are familiar with. As is evident, this technology is becoming the lighting technology of choice, not only for general home lighting but as the illumination source within every display technology. Thus it is important to understand this technology in a little more detail than we have discussed incandescent and fluorescent lighting.

With the invention of the first visible light iLED in 1962 and subsequent invention of high power yellow and blue iLEDs during the 1990s [5, 8, 12] these devices have been developed to have a number of characteristics which are desirable for lighting. iLED devices are comprised of a crystalline matrix of inorganic materials, attached to a pair of electrodes. The electrodes provide current at a specified voltage to the crystalline structure, injecting electrons into the crystal matrix. These electrons then travel through the crystalline structure until they become trapped at a defect within the structure. As the electron excites a molecule at this junction, the molecule eventually relaxes, resulting in the release of a photon. Through this process, electrical energy is converted to light. Because the materials within the inorganic iLED are highly controlled during manufacturing, most molecules within the device have a similar band gap, and therefore the photons all have a similar energy. Therefore, the light that is produced within an iLED is emitted with a narrow bandwidth, and thus have a narrow spectral emission. It is important that molecules which do not emit a photon through radiative relaxation dissipate this energy through spin-lattice relaxation, resulting in the release of heat. Ideal iLEDs thus produce near one photon for every electron injected into the device and produce little heat. However, most practical devices produce significantly less than one photon for every electron injected into the device and produce significant amounts of heat.

3.3.3.1 Electrical Properties

As the number of photons that are produced is proportional to the electrons injected into an iLED, the energy produced by the iLED is generally proportional to the current input to the device, as shown in Fig. 3.7. It might be useful that the heat produced within an iLED is also proportional to the current input to the device. As heat increases, the iLED can become less efficient. As a result, the amount of light produced for a given current is reduced. This reduction can result in downward curvature in the relationship depicted in Fig. 3.7 for higher current values. However, light output for an iLED is approximately proportional to the current input to the iLED.

The relationship between voltage and light output for an iLED is a little more complex than the relationship between current and light output. Similar to the performance of any diode, an iLED has a threshold voltage, below which no current will flow through the iLED and no light is produced. Once a threshold voltage is obtained, current begins to flow through the iLED resulting in light emission. Further increases in voltage then result in increasingly more current and, consequently, more light output. The relationship between voltage and current or light output from a typical iLED is shown in Fig. 3.8. Note that this relationship is highly nonlinear and can be approximated by a traditional diode function.

3.3 Characteristics of Artificial Lighting

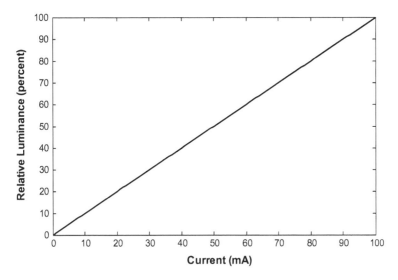

Fig. 3.7 Luminance output from an iLED as a function of current

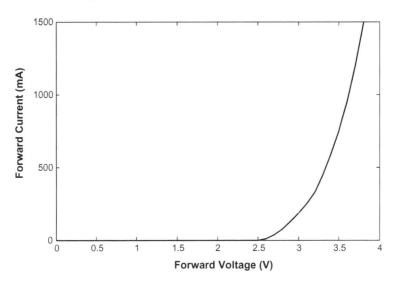

Fig. 3.8 Current as a function of voltage input to an iLED

3.3.3.2 Color Characteristics

Figure 3.9 shows the emission spectra for a typical red, amber, green, and blue iLED. As shown, the width of the spectral emission curve at 50% of its maximum energy is typically in the neighborhood of 30 nm for the red, amber, and blue iLEDs. These emission bands are especially narrow for long wavelength iLEDs. Green iLEDs are

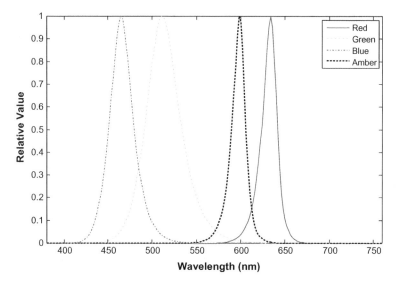

Fig. 3.9 Relative energy emitted by representative red, amber, green and blue iLEDs across the visible spectrum. Data adapted from Gilman and colleagues [2]

often the most challenging, often having broader emission spectra and lower absolute power than blue or red iLEDs. Because of their narrow bandwidth, iLEDs provide light that is relatively pure in color.

The color purity of the iLEDs, the spectral emission bands of which are shown in Fig. 3.9, is illustrated in the 1976 Uniform Chromaticity diagram shown in Fig. 3.10. As shown, the chromaticity coordinates of the red and amber iLEDs appear to lie on the spectral locus, having color purity that is perceived to be almost as good as a laser. The chromaticity coordinate for the blue iLED also lies very near the spectral locus. However, the chromaticity coordinates for the green iLED lies substantially inside the spectral locus due to the broader bandwidth of this emitter.

> LEDs are capable of providing a wide array of highly saturated colors. As a result, iLED lamps can be formed from several iLEDs and the designer has significant flexibility in forming the output color of the lamp.

Examining the color of light produced by an iLED, one would expect that an iLED could be produced to emit light centered on a desired wavelength. Therefore, the color of light produced would be highly controllable and stable. However, three primary issues arise in practical systems. First, the production of iLEDs is highly sensitive to variation, with variation resulting in differences in the spectral content of the emitted light. Secondly, the spectral output and the efficiency of the iLED can be sensitive to heat. The result is that the peak wavelength of light output from an iLED

3.3 Characteristics of Artificial Lighting

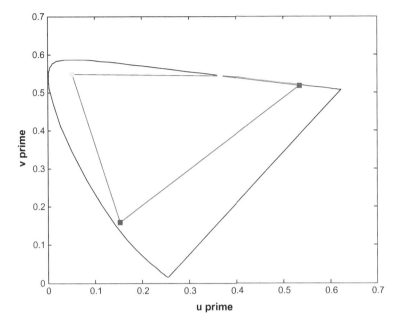

Fig. 3.10 Uniform chromaticity diagram illustrating chromaticity coordinates of representative red, amber, green, and blue iLEDs

can vary across each wafer that is produced and therefore even iLEDs produced on a single wafer during a single manufacturing run can vary by several nanometers, resulting in noticeable differences in color output. Thirdly, the color output a by a single iLED can vary by a few nanometers as the current to the iLED is increased and more heat is generated within the iLED. Once again, this variation can be large enough to be noticeable to the human eye. Finally, while highly efficient, high energy iLEDs are in production for producing blue, red, and infrared light; the production of highly efficient, green iLEDs continues to be a challenge. Similarly, iLEDs for producing very short wavelength ultraviolet energy is an area of intense research.

3.3.3.3 Drive Methods

Based upon the earlier discussion of electrical properties of iLEDs, it might be clear that one can control the current to an iLED to create a linear change in luminance or control the voltage to an iLED to create an exponential change in luminance output from the iLED. In effect, modifying either of the current or voltage, changes the current which flows through the iLED. As discussed, as the current through the iLED changes the color of light output from the iLED can also change. Therefore, it would be desirable to somehow drive the iLED such that when it is active, it is always driven with the same voltage and current, thus ideally always producing the same color of light.

To modulate the luminance of an iLED while driving it with the same voltage and current, one can modify the proportion of time that the iLED is turned on or off during the time over which the eye integrates. In this way, the human eye cannot detect the fact that the iLED is being turned on or off. However, because turning the device off a portion of the time reduces its time averaged output, it is perceived to be lower in luminance. This goal is usually achieved through pulse width modulation in which the time that the iLED is pulsed on or off is modified to change the time averaged luminance output of the iLED. Luminance thus changes linearly with the proportion of time the iLED is activated to the total possible time (e.g., turning the iLED on 50% of the time produces half the luminance the iLED would produce if were turned on 100% of the time.).

3.3.3.4 Spatial Distribution

ILEDs are typically packaged in a sealed container with the top of the container producing a crude lens, which shapes the spatial distribution of light from the iLED. The iLED itself is generally a Lambertian emitter. That is, an iLED will emit light in all directions with maximum intensity normal to the surface, which decreases in proportion to the cosine of the angle from the normal. However, iLEDs are normally constructed to be transparent only on one side, thus the light must pass through one side of the iLED. The lens is formed on the front of the iLED to protect the iLED from environmental oxygen and moisture, which can oxidize components, (e.g., the anode) within the iLED resulting in degradation. This encapsulation is often shaped to direct most of the light produced inside the iLED in a forward direction. Thus, the light is typically emitted within a focused cone from a small point source.

3.3.3.5 Color Conversion

Although iLEDs can be used by themselves, the presence of color variability, low efficiency green and cost often drives the use of color conversion materials when white light is required. As the blue iLED is very efficient, it is possible to produce blue iLEDs and then coat them with a phosphor or utilize them with optically-pumped quantum dots to create white light.

The most common approach, which is used in LCDs as well as general lighting is to form a mixture of phosphors which absorb a portion of the blue light emitted by the blue iLED and down converts this energy to longer wavelength yellow light. In this approach, either a broadband yellow or multiple phosphors can be mixed and coated on the iLED to create broad band emission. As only a portion of the blue light is absorbed by the phosphor and the phosphor emits yellow light, the iLED appears to produce white light emission. The resulting iLED will typically have a narrow blue emission peak and a broader yellow emission peak as shown in Fig. 3.11.

3.3 Characteristics of Artificial Lighting

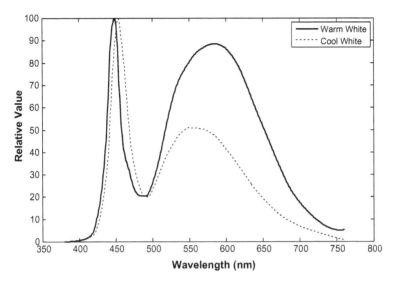

Fig. 3.11 Spectral emission of a pair of typical yellow phosphor-coated blue iLEDs for outputting white light with different correlated color temperatures. The dashed line represents a "cool", blue-shifted white while the solid line represents a warmer, yellow-shifted white

Notice that this figure shows a pair of curves for two phosphor converted iLEDs. An advantage of this approach is that different phosphors can be selected and mixed to have different bandwidth yellow phosphor emission. As the emission of these lamps is generally smooth and covers a large portion of the visible spectrum, these lamps can have relatively good color rendering and coloring rendering index values greater than 90 are frequently achieved.

> The availability of color change materials, including phosphors and quantum dots provide the designer even more flexibility in tuning the output color and spectra of iLED lamps.

Similarly white light can be produced using blue-emitting iLEDs in combination with appropriately sized quantum dots. Quantum dots typically convert a high energy photon to a lower energy photon, based upon the size of the quantum dot. By manufacturing quantum dots which are similar in size, narrow-band longer wavelength light can be produced. Light from blue iLEDs can then be used to excite quantum dots having similar sizes to produce narrow-band longer-wavelength light emission. Films can be produced which have two or more sizes of quantum dots to convert light to two or more narrow bandwidths. For example, films can be produced with two sizes of quantum dots. These films are then capable of converting blue light

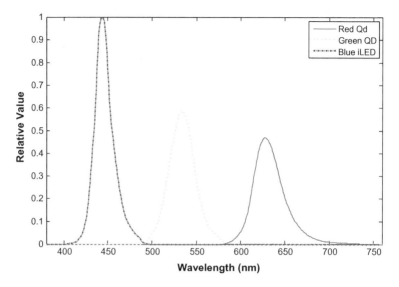

Fig. 3.12 Spectral emission for a blue iLED together with representative green and red quantum dots

into white light. However, the emission of energy from the quantum dots is narrow, emitting in the green and red portions of the visible spectrum as shown in Fig. 3.12.

Theoretically, the quantum dots may be coated directly on the iLED, however, they can be heat sensitive and are therefore typically coated on a separate film, removed from the iLED. Phosphor-coated iLEDs have been used in both general lighting applications, as well as LCD backlights. Quantum dot films are currently expensive and have not been used in general lighting, but are being used in high end LCD backlights. It is possible that mixtures of quantum dots could be coated on a film to produce broader-band emission. Further, research has shown that at least 4 emission peaks are required to reliably achieve a color rendering index with emitters that are 30 nm in width [9] and as many as 13 peaks are required to reliably mimic daylight spectra [10].

3.3.3.6 Additional Comments on iLEDs

The efficiency of iLEDs is high and becoming higher. Efficacy, the efficiency of light production, is often compared across technologies in units of lumens per Watt. Traditional incandescent light bulbs typically produce light with an efficacy of 8–17 L/W. Fluorescent bulbs, depending upon their design, often deliver light with an efficacy of 70–110 L/W. Unfortunately, when placed in realistic lighting fixtures, this efficacy drops significantly, often resulting in efficacies from 30 to 70 L/W. Today's iLED light fixtures can deliver efficacies of 70–110 L/W and fixtures have been demonstrated with efficacies of up to 200 L/W [13]. Therefore, iLEDs are

3.3 Characteristics of Artificial Lighting

beginning to replace the incumbent technologies because of this gain in efficiency. The importance of these efficiency gains is evident as the power consumed in the United States for lighting was approximately 641 TWh in 2015, or about 17% of total electric power use [1]. The transition to higher efficiency lighting is reducing power consumption as average system efficacy of installed lighting increased from 36 lm/W in 2001 to 51 lm/W in 2015 [1]. The transition to iLED-based lighting is likely to further improve system efficacy.

Thus, iLEDs are becoming a mature, mass-produced technology. They are formed on wafers that are 10s of cm in diameter and then divided and packaged into final devices which are 3–5 mm on each side. They can produce narrow-band, highly saturated light with among the highest efficiency of any light producing device in existence. These devices can then be combined with color conversion devices to tune the light output from the devices. Although highly efficient, the resulting devices are small and serve as highly directional point sources. Further, the exact colors of iLEDs vary in manufacturing and the exact color can vary as a function of their drive current. Finally, very highly efficient blue, red, and infrared iLEDs are currently being mass produced, while research continues on green and ultraviolet iLEDs.

As we have seen, iLEDs offer a lot of flexibility in constructing devices capable of outputting various colors of light. As a result, there is significant ongoing research to explore the use of iLED illumination devices which are capable of adjusting the color of white light they output similar to the way that the color of sunlight changes throughout the day [9]. Such devices hold the promise of creating artificial light which is more natural.

> Because of their low voltage drive characteristics and flexible color, iLED lamps can be created with adjustable color. Although, color adjustment might be used to provide many features, one can build lamps which change color throughout the day to mimic natural daylight.

3.4 Reflectance of Natural Objects

As we talk about illumination and color in the world, it is important to understand the characteristics of color in our natural world. Also, the relationship between color and brightness of these colors is important.

3.4.1 Color in Our Natural World

On earth, the sun provides energy to the surface. Living organisms absorb a significant amount of the sun's energy to support life. As an example, leaves of most plants absorb both blue and red light to support photosynthesis. Green light is reflected, giving the leaves we discussed in Chap. 1 their green appearance. To absorb the sun's energy, many objects absorb a significant amount of light across the entire visible spectrum, as such these objects reflect little light and appear dark.

Additionally, few natural objects absorb most wavelengths of light, but reflect narrow bands of energy. The light reflected by most objects is not highly saturated. Further, it is generally accepted that average luminance in a scene is approximately 20% of the white luminance [7]. Investigating the relationship between color saturation and luminance in specially selected high dynamic range pictures, it has been shown that the points in the image with the greatest color saturation have a luminance of 0.0316 times the scene luminance. As the luminance increases, the boundary of saturated colors generally decreases. Beyond 3.6 times the average scene luminance, the saturation of the colors decreases substantially [3]. This same research, however, shows that yellows with some color saturation can exist for some colors which have a higher luminance than white.

> Very saturated colors appear in the world, but are rare. Most of these saturated colors are dim, such that highly saturated and very bright colors are exceedingly rare.

What are the implications of this for an image system? First, half of the pixels in an image will generally be rendered to have a luminance that is only 20% of the white luminance. So most pixels in an image require only a small portion of the white luminance that a display can create. Secondly, most of these colors are not very saturated, that is they have chromaticity coordinates near the white point of the display. This does not imply that color is not important. Rendering the few very saturated colors from the natural world with saturated color is very important. We just need to recognize that relatively few pixels in any image are likely to have saturated colors. Finally, when these saturated colors do exist, they will generally be in the low luminance part of the image. That is, colors that are both highly saturated and bright are quite rare, although they do occur.

3.4.2 Relating Color Saturation and Reflected Power

It is important that there is a clear relationship between color saturation and reflected power. Generally, material within the natural environment absorb, reflect or transmit energy across a broad range of wavelengths within the electromagnetic spectrum.

3.4 Reflectance of Natural Objects

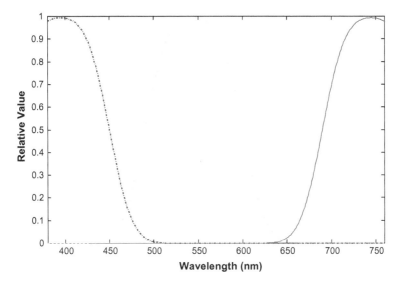

Fig. 3.13 Illustration of three broadband spectra for three imaginary colorants

Even most dyes or pigments, which are used as commercial colorants tend to reflect or transmit light across a broad range of wavelengths. To affect color, this range must include a portion of the visible spectrum. To create very saturated colors, it often becomes necessary to mix multiple dyes or pigments to create narrower bands of emission.

To understand this, let's begin by looking at three imaginary colorants. We will assume that these colorants transmit energy within a given broad band as shown in Fig. 3.13. As shown in the figure, each of these colorants transmit light over a similar width band (i.e., have a similar bandwidth) within the electromagnetic spectrum. Note, however, that the colorant on the left transmits a large portion of its energy at short wavelengths, most of which are lower than the lower bound of visual sensitivity (i.e., 380 nm). The colorant represented by the passband in the middle of Fig. 3.13 transmits all of its energy within the visible spectrum. Finally, the colorant represented by the passband on the right transmits wavelengths in the red portion of the visible spectrum but also passes a significant amount of infrared energy. How does the location of these passbands affect the amount of color saturation and luminance of these three colors? The answer to this question is provided by examining Table 3.1.

Table 3.1 shows the 1976 uniform chromaticity coordinates for the color from each of these colorants when illuminated with a light source which has equal energy across the visible spectrum, the calculated color saturation from Eq. 2.16, and the relative luminance with respect to white. As shown in this table, the colorants which overlap the edges of the visible spectrum are more saturated than the one in the center. This is probably not surprising. The overlap in their passbands and the visible

Table 3.1 Resulting uniformity chromaticity coordinates, saturation and relative luminance for emission through the filters in Fig. 3.13

Spectra	u'	v'	Saturation	Relative luminance
Short Wavelength	0.2228	0.0520	5.27	0.0037
Middle Wavelength	0.1523	0.4485	1.34	1.0000
Long Wavelength	0.6150	0.5078	5.48	0.0126

spectrum is narrower than the overlap in the passband and the visible spectrum for the centrally located spectra, therefore the color is more saturated. Additionally, the relative luminance of the colorants at the edges of the visible spectrum is also significantly lower than the centrally-located spectra. As most of the energy these colorants pass are in the ultraviolet or infrared portions of the visible spectrum, this is perhaps also not surprising. However, selecting passbands which only overlap the extremes of the visible spectrum, is one means of obtaining saturated color and this method obviously reduces the luminance of the saturated color.

Another method for creating a saturated color from a typical broad passband material is to use two materials where the passbands of these two materials partially overlap. Figure 3.14 provides an example where two of these passbands partially overlap. As we mix these materials together, each material absorbs a portion of the light with the spectrum of absorption defined by the passband of each material. If we shift these passbands so that they have less overlap, then the passband becomes narrower, resulting in greater color saturation. However, narrowing of the passband also reduces the relative luminance of the resulting color. Further, as the passband becomes narrower, its height or amplitude can be reduced, further reducing the relative luminance of the light which passes through the material.

Table 3.2 shows data similar to that shown in Table 3.1, where the center frequencies of the two passbands shown in Fig. 3.14 are separated further, providing less overlap. In this table, the first row indicates the spectra shown in 3.14. The following two rows show the same spectra with the exception that each pass band is shifted by 15 nm for each row. Once again, we see a clear relationship where increasing the color saturation also reduces the brightness (i.e., the relative luminance) of the color. This relationship exists for reflective or transmissive materials. Therefore, saturated colors in our natural world generally have a lower brightness than less saturated colors when illuminated similarly. This relationship then implies that saturated colors in our natural world will be dim compared to less saturated colors.

> Objects having highly saturated color generally reflect less light and will, therefore, be lower in luminance than less saturated objects.

3.4 Reflectance of Natural Objects

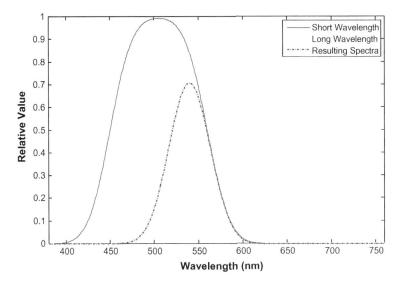

Fig. 3.14 Illustration of the effect of creating a color filter by mixing or overlapping a pair of color filters having the red and green spectra

Table 3.2 Uniform chromaticity coordinates, saturation, and relative luminance for the broadband spectra shown in Fig. 3.14 shifted to multiple center wavelengths, as shown

Wavelength	Wavelength	u′	v′	Saturation	Relative luminance
505	570	0.1040	0.5764	1.92	0.4310
490	585	0.0970	0.5795	2.02	0.1859
475	600	0.0931	0.5810	2.07	0.0446

Saturated colors are generally lower in brightness than less saturated colors when illuminated similarly. It is possible to provide more illumination to a saturated object or to have a saturated object reflect more light than a less saturated color. For example, if an object having a saturated color reflects most of its light in a given direction, this object can appear brighter than other objects around it. Additionally, if the saturated object is backlit, for example colored glass in a church window, the saturated color can be much brighter than other objects within the environment. The colors created under these conditions are visually very compelling as they only exist in very rare, specialized circumstances.

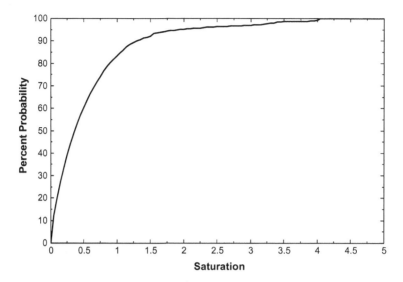

Fig. 3.15 Cumulative probability as a function of saturation for 3000 sample images

3.4.3 Color Occurrence in Images

Although characterization of natural images permits us to understand the average brightness and color saturation in natural images, it is also interesting to understand the distribution of colors in typical images. Previous research has shown that the average luminance is approximately 21% of the maximum luminance for television content [11]. This analysis also demonstrated that the majority of pixels are near neutral.

To provide a similar demonstration, 3000 pictures were downloaded from a free professional photography website [14] and subjected to analysis. In this analysis we assumed that each picture in the database was stored in sRGB format. These images were then decoded into a luminance chrominance space and saturation calculated for each pixel. A total of 1000 by 1500 pixels were selected from each image by determining the image size dividing the image height and width by the desired number of samples, determining a scaling ratio for the picture, scaling the 1000 by 1500 pixel sampling grid to the size of the picture and then selecting the closest pixel to each sample point, resulting in 4.5 billion samples.

Figure 3.15 then shows the cumulative probability of a pixel as a function of increasing saturation. As shown, approximately 20% of the pixels have saturation less than 0.2. As saturation increases, about 80% of the pixels have a saturation of 1 or less, 95% of the pixels have saturation less than 2, and 99% of the pixels have a saturation of 4 or less. As this figure indicates, the vast majority of pixels are low in saturation with only a small percentage of the pixels having a saturation greater than 2.

3.4 Reflectance of Natural Objects

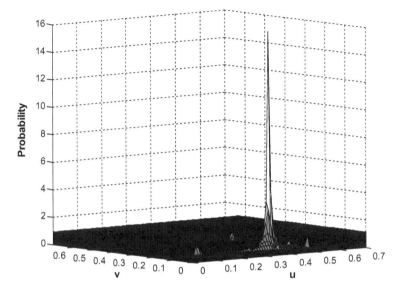

Fig. 3.16 Three-dimensional probability plot as a function of 1976 uniform chromaticity coordinates

We can also view this data in a chromaticity diagram. Figure 3.16 shows the probability of obtaining u'v' values. In this figure, u'v' has been binned into buckets of 0.01 units. We see that there is a spike in this probability function near neutral. In fact, approximately 16% of the pixels are within 0.01 u'v' units of the white point in the image. The probability then decreases as the distance from neutral increases. Probabilities associated with colors near the edges of the sRGB primaries are very low, in fact so low they are not rendered in this figure. However, there are small peaks in probability near each of the primary and secondary colors. Therefore, we can see that the vast majority of pixels within this sample of images are near neutral.

> The vast majority of the area in our natural world is low in saturation, having chromaticity coordinates near the white point of the scene, often referred to as neutral.

We can also plot saturation for various luminance levels. For this analysis we will divide our luminance space into 6 regions. The first five of these include luminance ranges of 0 to <20% of white, 20 to <40% of white, 40 to <60% of white, 60 to <80% of white, and 80–100% of white. Note that the sRGB standard specifies white at a code value of 235. Therefore, code values above 235 are reserved for highlights and colors brighter than white. The 6th region then will include colors brighter than white (i.e., greater than 100% of white). These six plots are shown in Fig. 3.17. As shown, the top three plots show the results for the 0–20, 20–40 and 40–60% luminance

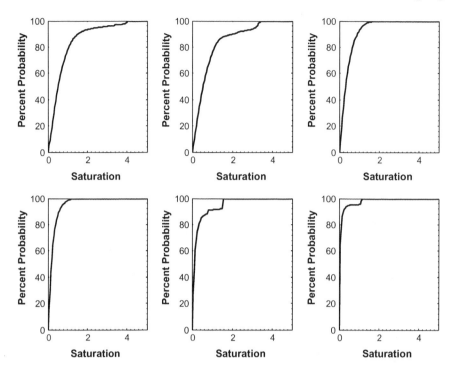

Fig. 3.17 Probability as a function of saturation for various luminance levels. Individual plots show the lowest luminance level in the top left with luminance increasing to the right. The highest luminance level is shown in the bottom right plot

conditions. The bottom three plots show the results for 60–80, 80–100% and greater than 100% of white conditions.

As shown in top left and top center plots of Fig. 3.17, saturation values of up to 4 are present when the luminance is less than 20% of white and saturation values up to 3.5 are present when the luminance is less than 40% of white. Beyond these luminance values, no pixels exist with saturation much beyond 1.5. This clearly illustrates that most saturated colors are low in luminance. In fact, most highly saturated colors are less than 40% of white within this particular image set. Note that as these images are professional images that the photographers have selected for sale, these photographs are generally well exposed and rendered. It is also worth noting that about 63% of the samples have a luminance less than 40% of white and about 7% of the samples have a luminance greater than 100% of white. Therefore, we can see that not only are saturated colors generally low in luminance, but most of the pixels are also low in luminance.

3.5 Summary and Questions for Reflection

In this chapter we reviewed methods for measuring luminance, which is a physical quantity which correlates with perceived brightness. We also discussed the measurement of illuminance, the amount of light hitting a surface. This measurement is important as illuminance reflected from an object becomes luminance. We then discussed lighting. We discovered that the color and intensity of sunlight varies throughout the day, but we perceive little change in the color of light due to adaptation. We saw that traditional artificial lights do not change in color and their intensity is generally not varied. However, artificial light formed from iLEDs hold promise for cost effective and relatively easy adjustment, making color and intensity adjustable artificial illumination much easier to attain. Examining how this light is reflected from objects in our environment, we saw that generally the relative luminance, often referred to as lightness, is often reduced as the saturation of reflected or filtered light is increased. This relationship exists in measurements taken from the natural world and in photographs. In these environments, not only are bright colors desaturated, but the preponderance of area within the natural environment is relatively desaturated. In later chapters, we will examine the implications of these relationships further, especially their implication for multi-primary displays. However, for now, here are some questions to consider as we progress further into the system.

1. Is the change which occurs in the color of the light throughout the day desirable to replicate in man made environments?
2. How important is it to replicate the change in luminance which occurs in natural light throughout a day?
3. If we seek to change the color or luminance of artificial illumination in our indoor environments, what changes should we make to our displays as this luminance changes?
4. Is it possible that the static nature of our indoor lighting is affecting our health by affecting our sleep and/or the release of hormones within our bodies?
5. If saturated colors occur only infrequently in the natural world, what are the implications for the importance of these colors to the human perceptual and cognitive systems?
6. If saturated colors are generally low luminance colors, how should this affect the design of our imaging systems?

References

1. Buccitelli N, Elliott C, Schober S, Yamada M (2017) 2015 U.S. lighting market characterization. Washington D.C
2. Gilman JM, Miller ME, Grimaila MR (2013) A simplified control system for a daylight-matched LED lamp. Light Res Technol 45:614–629
3. Heckaman RL, Fairchild MD (2009) Jones and Condit redux in high dynamic range and color. In: seventeenth color imaging conference, Albuquerque, NM, pp 8–14

4. Hernández-Andrés J, Romero J, Nieves JL, Lee RL (2001) Color and spectral analysis of daylight in southern Europe. J Opt Soc Am A 18(6):1325. http://doi.org/10.1364/JOSAA.18.001325
5. Holonyak N, Bevacqua SF (1962) Coherent (visible) light emission from Ga(As1-xPx) junctions. Appl Phys Lett 1:82. http://doi.org/http://doi.org/10.1063/1.1753706
6. International Commission on Illumination (CIE) (2004) Colorimetry (No. 15.3)
7. Jones LA, Condit HR (1941) The brightness scale of exterior scenes and the computation of correct photographic exposure. J Opt Soc Am 31(11):651–678
8. Kuoo CP, Fletcher RM, Osentowski TD, Lardizabal MC, Craford MG Robbins VM (1990) High performance AlGaInP visible light-emitting diodes. Appl Phys Lett 57:2937. http://doi.org/http://dx.doi.org/10.1063/1.103736
9. Miller ME, Gilman JM, Colombi JM (2016) A model for a two-source illuminant allowing daylight colour adjustment. Light Res Technol 48(2). http://doi.org/10.1177/1477153514559796
10. Miller ME, Madden TE, Cok RS, Kane PJ (2010) Lamp with adjustable color. United States Patent Number 7,759,854
11. Miller ME, Murdoch MJ, Ludwicki JE, Arnold AD (2006) Determining power consumption for emissive displays. In SID symposium digest of technical papers, San Francisco, CA, pp 482–485
12. Nakamura S, Senoh M, Mukai T (1993) High-power InGaN/GaN double-heterostructure violet light emitting diodes. Appl Phys Lett 62:2390. http://doi.org/http://dx.doi.org/10.1063/1.109374
13. U.S. Department of Energy, B.T.O. (2017) Caliper: snapshot downlights, Washington D.C
14. Unsplash: beautiful free photos (2017). Retrieved from https://unsplash.com/
15. Wyszecki G, Stiles WS (1982) Color science: concepts and methods, quantitative data and formulae (2nd ed). John Wiley & Sons, New York, NY

Chapter 4
Capture, Storage and Transmission Systems

4.1 Digital Capture Systems

Like digital display, digital capture systems are undergoing significant innovation. Digital cameras have traditionally been used to capture three-color images with limited dynamic range. In this chapter, we will primarily discuss this class of capture devices. It is important, however, to acknowledge that significant innovations are being made to develop very high dynamic range image capture, as well as to capture the distance of each object in a scene. This capture of distance provides the opportunity for novel image rendering and display.

4.1.1 Digital Camera Structure

The general structure of a digital camera, whether a standalone device or a component within a cellular telephone, is shown in Fig. 4.1. As shown, the typical digital camera consists of a lens to focus light on the image sensor. Many digital cameras additionally include some form of anti-aliasing filter to prevent spatial frequencies greater than can be registered by the sensor from reaching the digital sensor. A color filter is formed on the sensor to permit the capture of color information.

Importantly, cameras also have a number of components which control the amount of light that reaches the sensor. For example, most cameras will include a flash or a connection to an external flash, which permits the introduction of additional light under low lighting conditions. While a flash can provide a very intense light in the environment during the time that the digital sensor is receiving light, it can also demand significant power from the camera. Therefore, there is a need to limit the light output from the flash to limit battery drain. A flash can also introduce other unwanted artifacts, such as red-eye, unnatural shadows, and mixtures of colors of

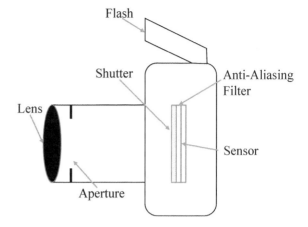

Fig. 4.1 Depiction of typical components in a digital camera

light in circumstances where the color of light provided by the flash differs from the light in the natural environment.

Typically cameras have an aperture to control the amount of light which reaches the sensor and the sensor is often equipped with an electronic shutter to control the amount of time that the sensor integrates light before permitting the light to be read out of the sensor. Each of these mechanisms control the amount of light which reaches the sensor with different effects. By closing the aperture, the amount of light entering the camera is reduced and the depth of field of the image is increased. By reducing the amount of time the shutter is open, the amount of light reaching the sensor is reduced. Long shutter times can cause the image of moving objects to be blurred as the image of the moving object moves across the sensor while the shutter is open. Short shutter times can reduce blur, but these short shutter times also reduce the amount of light available to create the image on the sensor.

The sensor may also have a gain, which amplifies the electric signal from the digital sensor as it converts photons to electrons. Other components which can be present are range finders, which determine the approximate distance to objects in the environment, and illumination sensors, which provide estimates of the light level within the environment. Of course, digital cameras also have digital processors to allow the image data to be manipulated and digital storage to permit images to be saved.

It is important that the flash, aperture, and electronic shutter can be used to modify the number of photons from the environment which are integrated to provide a digital signal from each pixel within the digital camera. Each pixel has an inherent noise level. Therefore, the larger the number of photons that enter a pixel in the digital camera, the higher the signal to noise ratio for that pixel. Having a high signal to noise ratio is important as the higher this quantity, the less likely the user will perceive noise in the final image. Therefore, these components may permit scenes with a large range of average luminance values to be captured by the sensor.

Importantly, a high quality digital capture system will control or provide light such that the full dynamic range of the sensor will be used to capture the scene regardless of the scene luminance. That is, the digital device will control the integration time of the cameras to a level that some pixels will receive enough photons to provide maximum exposure while other pixels will receive few if any photons. Changes in aperture, shutter time, and illumination permit the digital camera to control the exposure in response to differences in scene luminance using a process that is roughly analogous to the adaptation of the human visual system. Thus the digital camera captures scenes in a way that they are recognizable to the human viewer. However, this capture method practically eliminates all image information regarding the absolute luminance level between any two pictures, with the exception that images captured under low light may have very dark backgrounds and include higher levels of noise. Of course, if the aperture, shutter time and flash information is known, estimates of scene luminance can be calculated and stored with the image data.

Digital capture systems permit images to be captured over a large range of available illumination conditions, using flash to enhance scene lighting when sufficient light is not available. However, absolute luminance information is generally eliminated from the image data during the capture process.

4.1.2 Color Filter Design and Image Reconstruction

Like the cones in the human eye, color filters in digital cameras tend to have a broad bandwidth and have significant overlap. This characteristic is necessary as the broadband color filters absorb a larger portion of the light than narrow bandwidth color filters, giving the camera greater sensitivity in low light conditions. The use of broadband color filters also permits the pixels behind the color filter to be sensitive to all wavelengths of light within the visible spectrum. A color matrix is then applied to convert the captured electric signal for these three color filters to a different set of primaries, for instance primaries associated with XYZ tristimulus values or the primaries of a target display.

4.1.2.1 Bayer Color Filter Designs

Importantly, the arrangement of different colors of color filters on light sensing elements within a digital camera are typically not a repeating pattern of the three color filters arranged in a stripe pattern, as is common in many display designs. The most common arrangement of colors of light sensitive elements within a digital camera is the Bayer pattern shown in Fig. 4.2. This spatial pattern consists of a repeating array of a 4 element group, with two corners of this group having green sensitive elements and the remaining two corners having either red or blue sensitive elements. This arrangement of two green, one red, and one blue color filters is repeated across the sensor.

Note that this pattern contains twice as many green color-filtered light sensitive elements as red or blue color-filtered light sensitive elements. In fact, each row and

Fig. 4.2 A depiction of the Bayer color filter array applied in many digital cameras

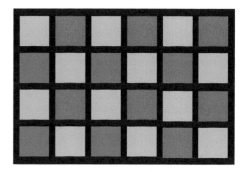

column of light sensitive elements includes green light sensitive elements while every other row and column includes red and blue sensitive elements. This arrangement, invented by Brice Bayer at Eastman Kodak Company, was designed with the recognition that the human visual system relies predominantly on luminance information for understanding image content [4] and the green sensitive elements have a response that permits them to receive a higher proportion of the luminance than the red or blue light sensitive elements. Therefore in a sensor having red, green and blue filtered light sensitive elements, the green filtered light sensitive elements carry much more of the luminance information than the red and blue filtered light sensitive elements.

Spatial image processing in a digital camera is also unique from the processing of spatial information in digital displays. The goal of digital image processing in a digital camera is to estimate the intensity of the colored elements where they are not present. That is, the red or blue intensity values are estimated at the location of each green light sensitive element. For instance, one might process each color channel to estimate the location of edges in the scene sampled by the sensor. This information can then be used to direct interpolation algorithms which effectively average digital information obtained for all surrounding light sensitive elements when no edges are present. However, when an edge is present, the algorithm might average digital signals from the light sensitive elements on one side of the edge to estimate the digital signal values for the light-sensitive elements [1]. As such, the algorithms attempt to reconstruct full resolution imagery, even though only one color is known exactly for each light sensitive element. Because there are fewer red and blue light sensitive elements than green, errors in the estimated values are typically larger for the red and blue digital values than for the green digital values. However, because the red and blue light sensitive elements provide signals which carry little luminance information, the errors in these color channels are often less visible to a human viewing the final image than similar errors in the green channel.

> The design of digital sensors and image processing involves the reconstruction of scene information from a partially sampled representation of the natural world.

4.1.2.2 RGBW Color Filter Designs

Although we discussed mechanisms in a digital camera which control the amount of light reaching the sensor, there is one more factor which influences the number of photons (i.e., the amount of light) that each light sensitive element receives. This factor is the size of the light sensitive element. The larger the light sensitive element, the more light it can gather. As the noise level in each light sensitive element is somewhat fixed, smaller light sensitive elements often have lower signal to noise ratios, resulting in more visible noise at low light exposure levels. This results in a significant trade-off as an increase in the resolution of a digital sensor typically increases the sharpness and, therefore the image quality, of the pictures that can be captured. However, as the resolution increases for a fixed sized sensor, the size of light-emitting elements decreases, and the signal to noise ratio decreases. Therefore, increasing the resolution of a sensor results in improvements in image quality for areas of an image in which the light sensitive elements receive plenty of light. However, the image quality of images captured with these same sensors decreases in the underexposed regions of images and for any underexposed images. While we associate increased image quality with higher resolution sensors, this relationship does not hold for all images due to this increase in relative noise level.

It is also important that as the resolution of an image increases, the resolution of the color information in an image can exceed the ability of our eye to perceive this color information. For very high resolution sensors, it is possible to add white light, unfiltered, or panchromatic light sensitive elements to a digital sensor [10]. By removing the color filters from some of the light-sensitive elements, the amount of light these sensors receive for a given set of conditions increase, increasing the sensitivity of the sensor and the resulting signal to noise ratio. Therefore, adding this fourth panchromatic light sensitive element has the potential to improve the image quality of images from high resolution image sensors. Although many arrangements of color-filtered light sensitive elements could be formed which include panchromatic light sensitive elements, a desirable arrangement might appear as shown in Fig. 4.3 [2].

Because the image processing techniques applied in digital cameras attempt to reconstruct the underlying image, the number of panchromatic light sensitive elements can exceed the number of red, green, and blue light sensitive elements. In the arrangement shown in Fig. 4.3, the arrangement of color filters is defined in each 3×3 arrangement of light sensitive elements. The pattern of color filters for these nine light sensitive elements is then repeated across the entire sensor. Each replication of these 9 light sensitive elements include one red, one green, one blue and six panchromatic light sensitive elements. Image processing routines rely on each light sensitive element to determine the pattern of luminance for each light sensitive element and the color is defined by interpolating the red, green, and blue signals [15]. As such, the luminance signal is very accurate with little noise and any errors exist with the color information.

Fig. 4.3 A depiction of a digital camera color filter array including panchromatic sensors in an RGB color filter array

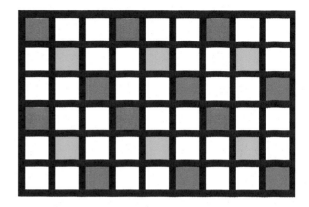

Panchromatic or white pixels have been used in digital cameras to improve the sensitivity of high resolution color sensors.

4.1.3 Image Integration

To fully understand the importance of panchromatic pixels in digital cameras and digital video cameras, it is useful to return to a discussion of the shutter. In a still scene, with nothing moving, it is possible for the digital camera to adjust the shutter time to capture images which are well exposed. Any image contains regions which reflect or produce more photons per unit of time than other regions. Ideally, images with reasonable dynamic range will have near white regions which provide some relatively large number of photons per unit of time at the sensor. The image will also have near black regions, which produce 2–3 orders of magnitude fewer photons per unit of time at the sensor than the near white regions. Under these conditions, the shutter can be programmed to stay open long enough that the total number of photons in the white region exposes the corresponding light sensitive elements to near their maximum value.

Note, however, that the number of photons per unit time changes as the scene luminance changes. In bright sunlight, the number of photons per unit time can be in the range of 50 million photons per second. However, under twilight conditions, the number of photons per unit time will be much smaller, perhaps in the range of 5000 photons per second. With all other variables equal, the shutter time for the twilight condition will need to be 10,000 times longer than the shutter time for the daylight condition. Therefore, if the shutter must remain open for 0.01 s to properly expose the sensor in bright light conditions, the shutter will need to remain open for 100 s to properly expose the sensor under low light conditions. If everything in the scene and the camera is perfectly still, then the shutter time can simply be increased. Unfortunately, many images are captured by individuals who hold cameras in their hand to capture a picture. As we all have a slight tremor as we attempt to remain still, we do not hold cameras perfectly still. Instead, we shake the camera slightly.

Further, many things in the scene may not remain perfectly still, leaves move in the breeze, people smile when they realize their pictures are being captured, and animals continue moving to gather food. Suddenly, a part of the image is moving on the sensor while the shutter is open and the sensor is integrating light. Now the image of the object is blurred as the object moves across multiple light sensitive elements while the shutter is open. To avoid this blur, the shutter time is reduced, the sensor receives fewer photons, the signal to noise ratio is reduced for most light sensitive elements and the image quality is reduced.

Under these conditions, the use of panchromatic light sensitive elements permit most light sensitive elements to receive more photons than light sensitive elements with red, green, or blue color filters. As such, the use of panchromatic light sensitive elements reduces the number of photons necessary to fully expose a light sensitive element, often by as much as a factor of 3. Thus, the use of panchromatic pixels can improve the signal to noise ratio of the sensor and improve the quality of images and particularly images captured under low light conditions.

> The use of four colors of light sensitive elements (e.g., red, green, blue, and unfiltered) increases the sensitivity of digital sensors, improving image quality under low light conditions when the sensor resolution is high enough to provide sufficient color information.

4.2 Image Encoding

In this chapter we have focused on the point that digital imaging systems typically capture images by manipulating light to be within a range that a sensor can capture and use specialized image processing techniques to reconstruct the pattern of light present in the natural image. An equally important system consideration for digital capture systems is that they must store the image information in a fashion which permits the image data to be decoded and displayed appropriately when viewed. As a result, digital file formats specify the format in which the image is stored.

Traditionally, imaging systems were designed with complimentary image capture, transmission and display formats. We can review the National Television System Committee (NTSC) standard or Phase Alternating Line (PAL) video systems as traditional imaging systems. In these systems, cameras were designed to produce video with a specific format. Storage and transmission systems were designed to store and transmit this specific format. Finally, Cathode Ray Tube (CRT)-based televisions were constructed to display video which they received in this format. The system was designed so that any component in the system was interchangeable and originally no component required significant signal processing. This was particularly true in the receiver and display portions, which were the only portions designed to be used by consumers. As such, the required signal processing was performed at the capture stage and images were saved in a format that permitted them to be easily decoded and displayed by low end consumer CRT televisions.

With the development of relatively low cost capture and digital signal processing, consumers now have devices which can readily capture, perform any desired image processing, store, decode the image with advanced digital processing and display the information. As such, many of today's systems can include digital image processing at each stage in the signal processing chain. That is, cameras, storage systems, transmission systems and displays each can include sophisticated digital processing systems to optimize their performance. However, the tradition of storing image data in formats which permit images to be easily displayed persists. Although there is no longer a single, dominant image format in any area of the world (e.g., NTSC or PAL) in which all electronic video or images are stored and transmitted, it is still the case that image encoding standards are developed which are intended to be compatible with displays having certain characteristics. Like the NTSC or PAL standards, the use of these image encoding formats can minimize the amount of processing that is performed by the display to create a desirable image. However, new data interchange standards are developed and displays must maintain backward compatibility. Therefore, displays or image decoding hardware must be capable of accepting images and video in several formats before rendering good quality images on the final display, regardless of the image encoding scheme and format.

> Traditional image encoding formats assumed that the image would be rendered during capture and production to be compatible with the desired output display. Originally this decreased the cost of electronic displays as this step eliminated the need for complex signal and image processing in the final display. With the advent of low cost digital image processing, the ability to provide custom digital image processing in the display provides a means for product differentiation, enabling displays to produce high quality images for many different input image formats.

Generally, the existing encoding schemes make a set of general assumptions about images and displays. The first of these assumptions is that the display will have a known white point and be calibrated such that the entire gray scale of the display is at this white point. As shown in Table 4.1, many of the current digital standards assume that the white point of the display resides at D65. This is a common assumption as it is believed that the visibility of images is improved when the display and surrounding lighting is matched in color [11, 19] and many viewing standards assume ambient illumination around 6500 K daylight [14]. Similarly, it is assumed that the white point of the image data should be rendered to match the designed white point of the display. Note that the camera must be able to estimate the white point of the illuminant in which the picture is captured and adjust this white point. Many scenes can have multiple illuminants, including artificial illumination, natural illumination entering the environment, perhaps through a window, and a flash, all of which can have different color. Under circumstances such as this, it is nearly impossible to adjust the white point of the image as different regions of the image have different

4.2 Image Encoding

Table 4.1 Chromaticity coordinates standardized by various display and motion picture organizations. Included are chromaticity coordinates from the International Electrotechnical Commission 61966-2-1:1999 standard (REC709), Society of Motion Picture and Television Engineers (SMPTE) Committee on Television Technologies Working Group on Studio Monitor Colorimetry Conrac Standard (SMPTE C), European Broadcasting Union (EBU) Tech 3320 standard (EBU Tech), and the Society of Motion Picture and Television Engineers Engineering Guideline 432-1:2010 (DCI-P3)

Color space	White point		Red		Green		Blue	
	x	y	x	y	x	y	x	y
REC709	0.3127	0.329	0.64	0.33	0.30	0.60	0.150	0.060
SMPTE C	0.3127	0.329	0.63	0.34	0.31	0.595	0.155	0.070
EBU Tech	0.3130	0.329	0.64	0.33	0.29	0.60	0.150	0.060
DCI-P3			0.68	0.32	0.265	0.690	0.150	0.060

white points. However, attempts are made to render the image so that its white point is near the desired encoding white point.

Further, as shown in Table 4.1, the display is assumed to have a standard set of primaries. These primaries vary from standard to standard. However, newer standards usually assume primaries with higher saturation as the saturation of primaries in consumer displays has generally improved over the past couple decades.

A separate, and often overlooked, assumption is that a perfectly diffusing white within the scene is to be rendered at a code value below the maximum grayscale code value (e.g., 235, 235, 235 rather than 255, 255, 255 in an 8 bit image). This assumption permits highlights and fluorescent objects, which can have luminance values higher than those of a diffuse white object within a scene, to be rendered as brighter than white. It should be noted that typically less than 10% of the code values are reserved for highlights, despite the fact that the highlights in natural images may be 2–3 orders of magnitude brighter than a perfectly white diffuser. Therefore, the relationship from scene luminance to rendered code value usually compresses the luminance range associated with highlights into a much smaller range.

Finally, there is often an assumption of the relationship between code value and luminance of the output display. This relationship is typically nonlinear, usually having a shape reminiscent of CRT displays to maintain backwards compatibility as practically all previous consumer display standards made this assumption [13]. This nonlinear function also permits a relatively efficient coding of luminance information as it compresses code values in the dark region of the image where the human visual system is most sensitive to luminance distortions. An example of the function used to transform camera code values (C_{linear}), which are linear with luminance, to nonlinear code values (C_S RGB) for storage is shown in Eqs. 4.1 and 4.2. The relationship shown produces code values according to the sRGB standard from input code values where the input code values are assumed to be linear with respect to desired display luminance [18]. Usually this desired display luminance is linear with scene luminance, often with the exception of highlight in-formation as noted earlier. Note that this function is the inverse of a typical CRT display code value to luminance function, often referred to as the gamma function. The term "a" in this equation is a constant, assumed to be 0.055.

$$C_{sRGB} = (1+a) * C_{linear}^{\frac{1}{2.4}} - a, \text{ if } C_{linear} > 0.0031308 \tag{4.1}$$

$$C_{sRGB} = 12.92 * C_{linear}, \text{ otherwise} \tag{4.2}$$

The CRT-shaped function is not applied in all standards, particularly some professional standards. For instance, the DICOM standards for medical radiology [3], assume a code value to luminance transfer function which is intended to represent observer sensitivity to changes in luminance. Such a function is designed to minimize perceptual distortions of luminance which can occur when the user is able to perceive luminance differences which are smaller than the minimum difference in luminance between neighboring code values. That is, this function is designed to have a shape which mimics the shape human's sensitivity to luminance differences.

Once the information is encoded, the image data typically undergoes some type of spatial image compression and is then stored in a standard image format. Ideally, image compression and decompression will have little effect on the resulting image or characteristics of the image important for image rendering. Generally, this step attempts to find and describe redundancy in the image information, removing this redundancy to minimize the size of the resulting image file. The file format can contain the actual image data but can contain additional information which might, under some circumstances, be useful during image rendering. For example, the file might contain information on the aperture setting, the shutter time, or other information which would permit one to estimate the absolute luminance in the scene which was captured.

4.3 High Dynamic Range

Another important attribute of imaging systems is their dynamic range, the range of luminance values that can be captured. Most traditional consumer digital cameras have the ability to capture light over a range where the brightest object is approximately 1000 times brighter than the darkest object in the scene. That is these devices have a dynamic range of approximately 3 log units. However, as we have discussed the natural light levels in the natural world often spans very large ranges, often 6 orders of magnitude or larger. This mismatch is handled by compressing luminance differences in the shadow and highlight regions of most images, condensing the light range down into 3 log units. Printed material and electronic displays have traditionally not been capable of presenting information over 3 log units of luminance.

4.3 High Dynamic Range

Therefore, the output of the display medium has limited the dynamic range of the system rather than the capture system.

Since, with the advent of LCDs with adjustable backlights and bright OLED displays, advanced display systems can present images with more than 3 log units of dynamic range. As a result, recent advances have been made in high dynamic range video capture.

Multiple techniques have been developed to improve the dynamic range of image capture systems [6]. Original techniques often involved capturing two or more images at different exposure levels, then combining these images to form a single image. The digital imaging technique to combine these images was derived from a manual method for reducing the dynamic range of high dynamic range images captured on film, a method referred to as "dodge and burn". In this technique, a high dynamic range image captured on film was transferred to a lower dynamic range medium, such as photographic paper, by effectively manipulating the time that the paper was exposed to light which passed through the film. The person printing film images onto paper basically inserted a light block between brightly lit areas of the film and the paper for a portion of the exposure time to reduce the intensity of light the paper was exposed to in these regions. This reduced the luminance range of the information and permitted the user to see detail in areas which would otherwise be clipped to black or white.

Modern Complementary Metal Oxide Semiconductor (CMOS) sensors provide the ability to address and read out data from select pixels, as well as the ability to read out the sensor at very rapid rates. As a result, many images can be read out of the sensor very rapidly and integrated to form very high dynamic range images. The resulting images can then be rendered to fill the dynamic range of the final display. As most displays available today are still limited in dynamic range, usually the dynamic range of these images is reduced during the rendering procedure. Using digital techniques similar to the dodge and burn technique applied in photography, the detail in areas of the image which are outside the dynamic range of the display are manipulated such that more of the detail is visible in the final display. While this technique improves the overall perceived quality of the resulting image, it does not preserve the relationship between scene and display luminance. Instead, high luminance areas of the image which require a luminance greater than the luminance range of the display are rendered with reduced luminance and dark areas of the image requiring a luminance less than the luminance range of the display are rendered with increased luminance. Therefore, this technique permits more of the information in the original scene to be displayed but it does not provide the same physiological sensation that we might perceive when viewing the natural image.

With the advent of improved displays, image formats are being standardized to maintain the original dynamic range of the image. Current standard file formats include Dolby Vision, HDR10 and Hybrid Log Gamma (HLG). The perseveration of high dynamic range digital information in a digital file format is not necessarily new as formats such as Photo YCC developed in the 1990s for perseveration of dynamic range from film scans and extensions to sRGB developed in the early 2000s accomplished similar goals [7, 17]. The current standards seek to provide methods of archiving high dynamic range images in a format which can be easily exploited when displaying these images on high dynamic range displays. Therefore, with well-designed displays, the current standards should permit high dynamic range scenes to be rendered on advanced displays such that they use the high dynamic range of the displays to more faithfully represent the changes in luminance within the original scene.

> The latest High Dynamic Range image formats attempt to provide high dynamic range video and images in formats which provide extended color gamut and very high resolution. Some of these standards aid the video producer in providing video which fulfills their intent when creating the content. Unfortunately, these standards often still assume a given response from the final video display.

4.4 Capturing the Third Dimension

While we generally consider image capture as a method for capturing a flat representation of the scene, several techniques are available for capturing depth information. In this section, I want to differentiate among three different techniques, including stereoscopic image capture, multi-view image capture, and depth capture. Each of these techniques have advantages and disadvantages, permitting the capture and sharing of more realistic experiences.

It is also important to differentiate the terms distance and depth. Distance typically refers to the length of the vector from oneself to an object in the environment. Depth refers to the length of the vector drawn between two objects in the environment where the vector is parallel to one's line of sight. Therefore, it is appropriate to discuss the distance from oneself to an object in the environment. Estimates of this entity are important in teleoperation when we are operating a vehicle or other object while viewing a display. Perception of depth (i.e., the distance between two objects in the plane parallel to one's line of sight) can also be useful for teleoperation but also provides other important perceptual cues which simplify other cognitive processes such as object segmentation. It is important that each capture technique provides depth information but only some of these techniques provide distance information.

4.4.1 Stereoscopic Image Capture

Beginning with the Wheatstone stereoscope in 1838, many photographic systems have been constructed in which a pair of images are captured of the same scene with

4.4 Capturing the Third Dimension

slightly different, horizontally-separated perspectives. When these two images are viewed simultaneously one by each of our eyes, we perceive a three-dimensional space. One can think of a stereoscopic capture system as a system which is mimicking the horizontal separation between our two eyes to present a pair of images, one image to each eye, which creates the perception of depth analogous to the viewing arrangement we experience under natural viewing. This model of stereoscopic capture is correct as a first approximation. Unfortunately, this concept of stereoscopic imaging as a system which mimics our eyes breaks down as the captured stereoscopic images are static while our eyes are constantly changing as they scan a scene. Therefore, stereoscopic imaging systems do not truly capture the scene as it would be viewed by our eyes, although stereoscopic capture systems capture images which are similar to an image that our eyes might capture at one moment in time when looking at a particular object within the scene.

To understand the similarities and differences between natural viewing and stereoscopic image capture, it is helpful to quickly compare the processes our eyes use when viewing a three-dimensional scene and the process used in stereoscopic capture. It is useful to make this comparison under two viewing conditions, first where the user is looking at a distant object and second where the user is looking at a close object. We will assume a standard image capture in comparison.

> Stereoscopic imaging involves the capture of a pair of images intended to mimic the two images our eyes capture of a scene. However, because our eyes change significantly as we look from object to object, these images fall short from replicating reality.

Figure 4.4 depicts human viewing of a distant object in the left panel, and a near object in the right panel. The distant object we will assume is near optical infinity. The near object we will assume to be relatively near, perhaps as close as 1 m away from the user. To understand human viewing, we must recognize that when viewing an object naturally, our two eyes rotate to permit the object to be presented on our fovea. In the left image, because the object is far away, the optical paths of our two eyes are nearly parallel to one another. However, when the object is near, our eyes rotate so that they converge upon the object and the optical paths of our two eyes converge on the object of interest. Importantly, our eyes do not rotate independently but rotate together under control of neurons in the midbrain when performing a vergence eye movement. Physically, the lens of our eye also focuses on the object of interest when this object is outside the eye's depth of field, in this example, focusing at optical infinity for the distant object and at 1 m for the near object. For many people, their accommodative response, which drives their focus distance, is highly interdependent upon their convergence distance, so the convergence and focus distance. As such, disparity and accommodation distance each appear to influence the perception of depth within a scene [8].

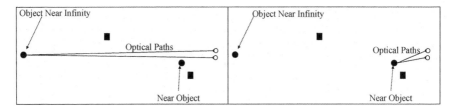

Fig. 4.4 Illustration of eye behavior when looking at a distant object (left panel) and a near object (right panel)

Notice also that there are other objects within the scene and these objects are now imaged at different spatial locations on the user's retina. For individuals with normal stereoscopic vision, the difference in retinal location is used by the eye-brain system to estimate depth between objects in the scene. We will rely on disparity to understand the depth between objects primarily when these objects are within or just beyond the eye's depth of field, so the angular separation of these objects from one another within the real world is often relatively small as compared to the possible range of disparities. Further, we can rely on other cues like shadows, shading, and size differences of objects to help us determine depth and distance. However, parallax cues, such as those derived from disparity, can localize relative depth differences more precisely than these pictorial depth cues.

Also note that if both the distant object in the left example and the near object in the right image were in the same scene, we probably could not see both objects clearly at the same time. The depth of field of our eye is likely such that when we look at one of these two objects, the other object is outside the range of distances that are in focus or nearly in focus. Therefore, if we wish to view these two objects, we will look at one object, center the object in our fovea by rotating our head to center the object, converge our eyes on the object and then focus on that object. To look at the other object, we will rotate our head to center the second object in our field of view, our eyes will rotate to center the second object on each of our fovea and the shape of our lens will change to focus the second object on our fovea. It is also important to recognize that when we look at the close object, if the object has depth, our two eyes will each see a different view of that object (the right eye will see more of the right side of the object and the left eye will see more of the left side of the object). When performing stereoscopic image capture of a scene containing both the distant and near object, we have the option of aiming our cameras by rotating them to a convergence point and focusing on a single object. However, in our example, the other of the objects are likely to be far outside the depth of field of the cameras and therefore the user will not be able to see both objects clearly.

Instead, we will likely point and focus our camera to some intermediate point as shown in Fig. 4.5. The use of this intermediate point permits both objects to be within the camera's depth of field, permitting both objects to be in focus. If the user then views this image on a stereoscopic display, the entire image is in focus. This is important as it permits the user to move their eyes around the scene and see each

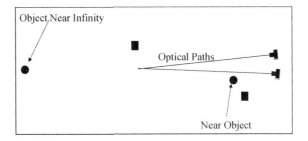

Fig. 4.5 Illustration of a typical camera arrangement used to capture the scene depicted in Fig. 4.4

object in the scene as in focus. However, regardless of the distance of the object, the user always focuses at the screen. Therefore, the consistency of the convergence point and the focus distance to objects in the scene no longer exists, which can produce eye strain for individuals, especially when their eyes seek to converge and focus at the same distance. Further, it is possible for the user to see the distant object when looking at the near object. Under this circumstance, the distant object now has a large angular separation on the human's two retinas and the human eye brain system may not perceive them as the same object, leading to further discomfort [9]. Further, unless the angular size of the scene and the display are maintained, these angular sizes change, reducing the utility of spatial cues, such as size constancy and the visibility of textures for distance and depth estimation. Additionally, although the user does see different views of the near object in their left and right eyes, these views are not geometrically the same as they would have been under natural viewing conditions.

Overall, stereoscopic image capture is the process of capturing two images of the same scene with cameras that are separated horizontally to mimic the separation of the human eyes. This process permits us to capture a pair of images which have a number of similarities to those captured by our eyes when viewing a three-dimensional world. Two slightly different views of our world are captured, in which objects at different depths cast images in the two eyes which are separated by a horizontal distance. This image separation between the two eyes permits us to perceive the relative depth among objects out in the visual world when we view these two images properly. However, the process does not permit the capture of the scene in a way that permits the focus distance of objects in the scene to change as we view different objects in the scene. Further, if we attempt to move our head when viewing the scene, we only have two representations of the scene available and the image is static as we move our head.

4.4.2 Multi-view Image Capture

Although a pair of cameras are required to capture stereoscopic images, this concept can be expanded to multiple cameras as shown in Fig. 4.6. In such a configuration, a

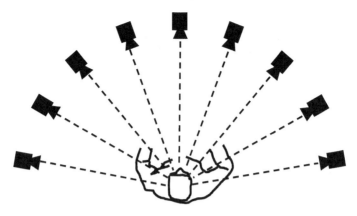

Fig. 4.6 Illustration of a multi-view camera system for capturing an image of a participant adapted from McDuff and colleages [12]

number of views of a scene can be captured and views can be taken from two or more of the cameras to reconstruct the scene when captured from an alternate viewpoint. This system has many of the same issues that stereoscopic imaging systems have. However, the presence of multiple views permits the scene to be updated or shown from multiple points of view as the participant moves their head or when the scene is viewed from alternate angles.

> Increasing the number of views increases the number of viewpoints which can be assumed when viewing stereoscopic images. However, the resulting images suffer from many of the same restrictions as stereoscopic images.

4.4.3 Depth Capture

Besides capturing multiple views of scenes to permit a user to perceive depth from stereoscopic images, several techniques are available for capturing the distance of objects from a camera. This can be accomplished by determining the disparity in object location within two or more scenes [5]. However, it often involves novel hardware, which permits distance information to be determined directly. This hardware can involve scanning laser systems where the duration from the time a laser illuminates a scene until the light returns is used to determine the distance to objects or systems in which the entire scene is illuminated and the time of flight is determined for each location in the image to determine the distance to objects [16]. Each of these techniques often require both an image sensor and a distance sensor. The distance map from the distance sensor is merged with the digital image to provide an image with a depth map. More recently, other approaches have been developed in which a single sensor can be used to both capture an image as well as to estimate the distance to each point within the image [20]. Knowing the distance to each pixel in a scene can permit digital manipulation of a scene to change the focus of objects at arbitrary distances within a scene. Such methods can provide the ability to render an image with different apparent depths of field, providing flexible image rendering.

> The addition of depth capture provides an estimate of distance to each object or pixel in an image, improving the ability to segment objects in the scene and render these objects differently depending upon where the observer looks within a scene. Combining stereoscopic or Multiview images with depth capture creates the opportunity to render scenes where the viewing characteristics of objects differ as we fixate upon them, creating a more natural experience.

4.5 Summary and Questions for Thought

We have discussed the fact that image capture systems control the amount of light interacting with a sensor to provide proper stimulation of the sensor. The use of auxiliary light sources and light control mechanisms allows the capture of light over a broad range of illumination levels on a sensor with a narrow response range. Therefore, much of the absolute luminance information in the natural scene is lost, much like it is lost while we look at the scene. However, there are distinct differences in the performance of capture systems and our eyes in terms of light capture. A significant one of these differences is that cameras capture entire scenes, primarily adjusting for average scene luminance and color of illumination while our eyes adapt constantly as we look around a scene.

We have also seen that digital cameras sample color and attempt to reconstruct the spatial information in the underlying scene based upon this sampled information. Traditionally, sensors applied three colored light sensors, however, with increasing resolution and the need for improved sensitivity, the use of unfiltered sensors together with RGB sensors has been developed and shown to provide significant improvement. We have also discussed the fact that the image information is then encoded, primarily to permit the images to be compatible with assumed displays. This system architecture was originally adopted to reduce the cost of consumer-owned hardware, however, we are in a transition phase where low cost electronics have enabled much of this processing to be shifted from the image capture to the image display system. Finally, we discussed some advanced capture features which are enabling improvements in realism of displayed images, including the use of high dynamic range capture as well as depth or distance capture. We will discuss the use of these image features to enable higher quality displays a little later in this book.

With this as a backdrop, let's consider some questions for thought.

(1) We have discussed how aperture and shutter speed are used to adjust sensor illumination during still image capture. How does this process change when capturing video images?
(2) If much of the luminance information is lost during image capture and the resulting images are viewed in a different viewing environment, is it possible to have similar experiences during image review as one would have had if they were viewing the original scene?
(3) If a panchromatic pixel can be added to an image sensor to improve sensitivity, what effect would adding a similar white pixel to a display have?
(4) If we were to design the image capture and storage system to permit the display the maximum flexibility when presenting images or scenes to a user, what image would we need to capture and how should we represent this information?
(5) To date, image capture systems have been developed with the thought that there was little diversity between displays. How does this approach affect the display industry and could a change in this approach affect the display industry?

References

1. Adams JE Jr, Hamilton JF Jr (2001) Adaptive color plane interpolation in single sensor color electronic camera. United States Patent Number 5,652,621
2. Adams JE Jr, Kumar M, Pillman BH, Hamilton JA (2012) Four-channel color filter array pattern. United States Patent Number 8,203,633
3. Association National Electrical Manufacturers (2011) Digital imaging and communications in medicine (DICOM) part 14 : grayscale standard display function. Rosslyn, VA. http://www.ncbi.nlm.nih.gov/pubmed/2188123
4. Bayer BE (1976) Color imaging array. United States Patent Number 3,971,065
5. Cremers D, Pock T, Kolev K, Chambolle A (2011) Convex relaxation techniques for segmentation, stereo and multiview reconstruction. In: Markov random fields for vision and image processing. MIT Press, Cambridge, MA
6. Darmont A (2013) High dynamic range imaging: sensors and architectures. SPIE Press
7. Giorgianni E, Madden TE (1998) Digital color management: encoding solutions. Addison-Wesley Longman, Inc., Reading, MA
8. Gooding L, Miller ME, Moore J, Kim S-H (1991) Effect of viewing distance and disparity on perceived depth. In: Merritt JO, Fisher SS (eds) Society of photo-optical and instrumentation engineers proceedings: stereoscopic displays and applications II, vol 1457, pp 259–266. http://dx.doi.org/10.1117/12.46314
9. Jin EW, Miller ME, Endrikhovski S, Cerosaletti CD (2005) Creating a comfortable stereoscopic viewing experience: effects of viewing distance and field of view on fusional range. In: Woods AJ, Bolas MT, Meritt JO, McDowall IE (eds) Society of photo-optical and instrumentation engineers proceedings: stereoscopic displays and virtual reality systems XII, vol 5664, pp 5612–5664. http://dx.doi.org/10.1117/12.585992
10. Kijima T, Nakamura H, Compton JT, Hamilton JF Jr (2010) Image sensor with improved light sensitivity. United States Patent Number 7,688,368
11. Lu W, Xu H, Luo MR (2012) Evaluation of contrast metrics for liquid-crystal displays under different viewing conditions. J Soc Inform Display 20(5):259–265

References

12. McDuff DJ, Blackford EB, Estepp JR (2017) The impact of video compression on remote cardiac pulse measurement using imaging photoplethysmography. In: 12th IEEE international conference on automatic face and gesture recognition. Washington, D.C
13. Miller ME, Yang J (2003) Visual display characterization using temporally modulated patterns. J Soc Inform Disp 11(1):183–190. https://doi.org/10.1889/1.1830239
14. National Geospatial Intelligence Agency (2012) NGA softcopy exploitation facility standard
15. O'Brien M, Pillman BH, Hamilton JF Jr, Enge AD, DeWeese TE (2010) Processing images having color and panchromatic pixels. United States Patent Number 7,769,229
16. Schuon S, Theobalt C, Davis J, Thrun S (2008) High-quality scanning using time-of-flight super resolution. In: Anchorage A (ed) IEEE computer society conference on computer vision and pattern recognition
17. Spaulding KE, Woolfe GJ, Joshi RL (2003) Using a residual image to extend the color gamut and dynamic range of an sRGB image. In: Proceedings of IS&T PICS conference, pp 307–314. http://scholar.google.co.uk/scholar?hl=en&q=Using+a+residual+image+to+extend+the+color+gamut+and+dynamic+range+of+an+sRGB+image&btnG=&as_sdt=1%2C5&as_sdtp=#0
18. Stokes M, Anderson M, Chandrasekar S, Motta R (1996) A standard default color space for the internet—sRGB. https://www.w3.org/Graphics/Color/sRGB
19. Tseng F-Y, Chao C-J, Feng W-Y, Hwang S-L (2010) Assessment of human color discrimination based on illuminant color, ambient illumination and screen background color for visual display terminal workers. Ind Health 48(4):438–446. https://doi.org/10.2486/indhealth.MS1009
20. Wang A, Gill PR, Molnar A (2011) An angle-sensitive CMOS imager for single-sensor 3D. In: IEEE international solid-state circuits conference, digest of technical papers. San Francisco, CA

Chapter 5
LCD Display Technology

5.1 Display Technology Overview

In previous chapters we have discussed a number of system influences on the perception of color in natural environments, as well as capture and transmission systems. These chapters provided the context in which display systems must be constructed and operated. We should, however, pause to acknowledge that the previous chapters have been image-centric. We expect displays to present images with high quality. However, we also expect high quality text and graphics, which we have not addressed. Generally, it is accepted that displays should be constructed to display high quality images and that users will accept the resulting text and graphics. Therefore, each of the previously-discussed system elements influences the requirements for a visual display within this larger environment. With this context in mind, we can now turn our attention to the display.

This chapter and the subsequent chapter provide an overview of three different technologies that are likely to influence displays into the future. We begin by discussing Liquid Crystal Displays (LCDs) in the current chapter. This is the dominant technology within the display industry. LCDs modulate light in response to an image signal. The base liquid crystal display technology does not emit light but instead modulates or controls the passage of light from a light source, referred to as a backlight, to the human observer. While fluorescent lights have traditionally provided a light source for LCD backlights, iLEDs are quickly becoming the light source of choice for LCDs. In traditional LCDs, the backlight provides a very bright uniform illumination and the liquid crystal layer permits only a small portion of this energy to pass through to the observer. As a result, the vast majority of the light produced by the backlight is absorbed by the display reducing the overall efficiency observed when converting electrons to photons which are available to the human eye.

Once we have reviewed LCD technology in the current chapter, we will turn our attention to Organic Light Emitting Diode (OLED) Display technology in the subsequent chapter. These displays are formed from an array of light-emitting diodes.

Therefore, each light emitting element in these displays emit light directly, much like the iLEDs we discussed in Chap. 2. However, the OLEDs are formed from organic-based materials which are coated onto a substrate. The substrate then controls the flow of current to each OLED within the display. This technology has the advantage that the display technology modulates the current provided to the OLEDs which emit light directly. Therefore, OLEDs modulate current, effectively modulating the intensity of the light which is produced in response to an image signal. As a result, electrons are only used to produce photons when they are needed. Therefore, similar to iLEDs in the lighting domain, OLEDs should be dramatically more power efficient than existing technologies. For this reason, OLED is quickly becoming the dominant technology in cellular telephones and has more recently been applied in televisions.

We will then introduce iLED displays in the subsequent chapter. These displays are formed from an array of inorganic light-emitting diodes. This display technology is not being mass produced and must overcome a number of significant technological barriers before it is ready to enter mass production. iLED displays do have some desirable properties which may draw significant long term investment. As such, these devices might hold promise in some future markets.

5.2 Liquid Crystal Display (LCD) Technology Introduction

The electrically switchable liquid crystal cell, which became the basis of LCDs, was first described by Frederiks in 1927 [1]. By the 1930s, the liquid crystal light valve was patented [2]. The first practical light-emitting display technology was invented in 1970 [4]. These devices were originally marketed as small, passive-matrix, reflective, monochrome devices. As such, they could be included in portable electronic devices, such as watches. Because they were small and consumed very little power, these reflective displays provided the ability to present information on devices that the Cathode Ray Tube (CRT) or Plasma displays, which were present in the market, could not. Therefore, they entered the market virtually without competition within these niche markets.

The LCD enabled the portable computer, such as the IBM Thinkpad 700C in the 1990s. These devices employed higher quality, backlit, pixelated LCDs and provided a market for new color LCDs. The original laptops employed VGA resolution with a diagonal dimension of around 10" and were less than 1" thick. Because it was not physically possible to create thin CRTs or color Plasma displays with this resolution, LCD technology again found a market without significant competition. As the portable computer grew in popularity, LCD manufacturers quickly built manufacturing capacity and improved the quality of this display technology. This technology then enabled image displays on portable devices, such as digital cameras and cellular telephones. As the basic building blocks could be scaled to both larger and smaller displays, this technology migrated to desktop computer screens and televisions where it was able to compete successfully with CRTs and Plasma. By 2010,

5.2 Liquid Crystal Display (LCD) Technology Introduction

the LCD had replaced practically all other display technologies in practically every display application.

> One of the primary differentiating features of the IBM Thinkpad 700 series laptops as compared to models from Compaq and others was the use of a 10.4" LCD as opposed to the 9.5" LCD. With this feature and the newly marketed Trackpoint device, customer demand outstripped production for many months.

Improvements in LCD technology required the improvement in several building block technologies, most of which were developed by different companies who participated within the LCD supply chain. This attribute of LCDs provided both a significant advantage and a potential disadvantage for the technology. The reliance of this display technology on several building block technologies was an advantage as the market developed. This structure permitted several companies to innovate, creating rapid improvements in the technology. That is advances were made in the light source and optical films within the backlight. These advances were relatively independent of improvements in drive technology, liquid crystal materials or optical films in the liquid crystal display layer. These advances were relatively independent of improvements in color filters and optical films which were necessary on the viewing side of the LCD. As such, advancements occurred rapidly as multiple companies participated in each of these technologies and competition between companies attempting to construct each component technology quickly drove improvements in image quality and reductions in cost. Like most markets, as this market matures, cost reduction becomes more important than quality improvement, prices drop, margins decrease, and the industry consolidates. With consolidation, innovation slows and different players in the value chain compete for remaining margin. As LCD requires a significant number of components, this competition restricts the ability of the industry to adopt lower cost structures. Never the less, this technology currently remains the dominant display technology.

> Modern direct view LCDs employ components designed and manufactured by several companies. This not only increases the rate of innovation but divides the cost of manufacturing infrastructure among several players, reducing the risk to any one company.

As we review this technology, we will begin by examining the overall structure of a typical color LCD. As we review the structure, we will discuss the primary function of each layer. After we review the basic structure, we will spend some time discussing some of the key performance characteristics of the LCD. We will then discuss areas of innovation, specifically we will focus our discussion on the use of enhanced backlights.

5.3 Technology Overview

The basic components of the color LCD are shown in Fig. 5.1. As shown we can divide the display into the backlight, the light modulator, and the light control layer. The backlight produces a very bright, typically very uniform white light source. The light modulator permits a portion of the light from the backlight to pass through select light emitting elements, permitting the creation of a monochrome image. The light control layer includes a color filter array that then converts the monochrome image to a color image and controls the reflection of light from outside of the display. Each of these three major assemblies is composed of multiple components.

The backlight is often formed from a reflector, a light source, a diffuser, and a light director. The reflector is typically a reflective surface that directs any light which reaches it towards the LCD panel. Traditionally, the light source was an array of fluorescent bulbs. However, modern LCDs typically include an array or matrix of LEDs. The array of LEDs could include either white LEDs, arrays of red, green, and blue LEDs, or an array of blue LEDs with a light conversion layer. As the light is emitted by a sparse array of fluorescent bulbs or LEDs, the light from these emitters must be diffused to create a uniform light source. Therefore, a diffusing film is used

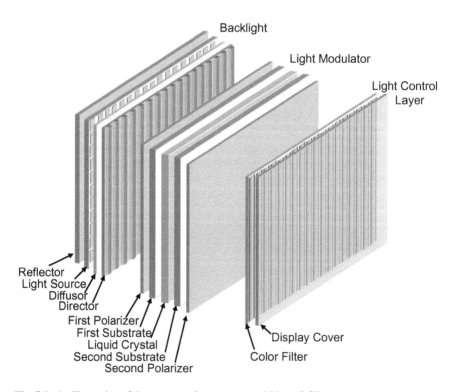

Fig. 5.1 An illustration of the conceptual components within an LCD

5.3 Technology Overview

to spread the light from the light emitters to create a uniform light source. Light emitted from the diffuser is typically released such that light rays are traveling in all directions. Therefore, one or more additional films are used to collect this light and direct it to travel in a path that is perpendicular to the light modulator. As noted earlier, the light source is provided by one of a group of potential suppliers. Additionally, multiple suppliers provide competitive sets of diffusing and light directing films. Additional suppliers integrate these components to form backlight components.

> The LCD does not create light at each light emitting element. Instead light is created everywhere and each light emitting element only blocks or permits some portion of the light to pass through the light modulator to be viewed by the user. As a result, the power consumption of the display is constant regardless of whether the display presents a black or a white screen.

Conceptually, the light modulator then permits a portion of the light from the backlight to pass through each light emitting element in response to an input image signal. The light modulator includes a polarizing film, a substrate having an array of electrodes formed on it, a layer of liquid crystal material, a second substrate having complementary electrodes and a second polarizing layer.

The concept of polarization is important in LCDs so it is useful to briefly discus this concept. As we know, light can be conceived as traveling in waves or as particles. Considering light as waves, these waves can be oriented in any number of directions. The ocean is covered by waves which travel horizontally across its surface. Similarly light can have waves which travel horizontally. However, unlike the ocean, light waves can travel through air in any orientation. They may be horizontal, like the waves on the ocean, vertical or any angle in between. In fact, the waves can even travel in a circular pattern. Natural light lacks polarization and therefore, it is composed of numerous waves where some of these waves are horizontal, some are vertical, and some are at every angle between these. Conversely, polarized light contains waves having only a single orientation, much like the waves on the ocean. Once light is polarized it has characteristics which can be more easily manipulated. It is also important that manipulation of the polarization of light changes the orientation of the wave but does not influence the frequency or wavelength of the wave. Therefore, polarization can be manipulated without changing the wavelength of the light which we associate with color changes.

Returning to our display structure, the polarizer in the light modulating layer receives polarized light. It then passes light having either a horizontal or vertical polarization while absorbing or reflecting the remaining light. An electric field is created between the pair of electrodes which is dependent upon the image signal. This electric field then controls the alignment of molecules within the liquid crystal layer. As the alignment of the molecules is adjusted, this layer can change the polarization of the incoming light (i.e., change the orientation of the wave). Using a twisted-pneumatic liquid crystal technology as an example, light emitting elements which are not supposed to emit light do not affect the polarization of the light while light emitting elements which are supposed to emit the maximum amount of light rotates the wave to the complimentary polarization. For example, the liquid crystal layer permits horizontally polarized light to be converted to vertically polarized light. Light emitting elements which are to emit intermediate amounts of light have their polarization partially rotated from one polarization to another. The light passing through the light modulator then encounters the second polarizing layer which permits light having a polarization perpendicular to the first polarizing layer to pass, while absorbing the remaining light. As such, only the light having polarization which was rotated by the liquid crystal material passes through the light modulator layer. As a result, this layer permits light to pass through some light emitting elements while allowing very little light to pass through other light emitting elements.

The light control layer then produces light with the appropriate color as it leaves the display and helps to reduce the reflection of light from the panel. This conceptual structure includes a color filter array. Typically this array includes repeating stripes of red, green, and blue color filter material. In the design of this layer, there is typically a significant tradeoff between the saturation of the color and the efficiency of light emission. As we discussed color materials in Chap. 3, the narrower the emission band of the color filter, the less light that can pass through the filter. This is due to the fact that both the bandwidth of the color filter and the peak emission amplitude are reduced as the bandwidth of the color filter is decreased. Of course, color saturation increases as the bandwidth of the color filter decreases, resulting in this tradeoff. Note that the color filters and the polarizer generally absorb a significant amount of light that passes from the backlight towards the observer but it also absorbs a significant amount of the light that enters the display from the ambient environment. Therefore, the LCD reflects little of the light from the environment which enters the display. However, the front of the display is formed from glass. This layer has a significant amount of specular reflection, similar to a mirror. An additional film can then be applied to the front of the LCD to absorb or diffuse the light which encounters the front of the display.

In general, these layers then form a typical backlit LCD. The layer structure shown in Fig. 5.1 is conceptual and layers can be reordered and in some cases multiple layers can be formed on a single substrate. Nevertheless, these basic components and functions are required in any modern backlit, active-matrix LCD. Further discussion of LCD types and more details on LCD function or structure have been provided elsewhere [7].

To understand the basics of the technology, it is important to return to the heart of the light modulator. A modern active matrix LCD employs a substrate containing a structure similar to the one shown in Fig. 5.2. As shown in this figure, the substrate contains an electrode. As the voltage to this electrode is varied, relative to an electrode on the second substrate, an electric field is created across the liquid crystal layer. The alignment of the molecules within the liquid crystal layer is controlled by the strength of the electric field. The voltage at the electrode is controlled by a voltage provided at a capacitor. This capacitor is charged by the flow of current through a Thin Film Transistor (TFT). The TFT permits a voltage signal which is provided on a data line

5.3 Technology Overview

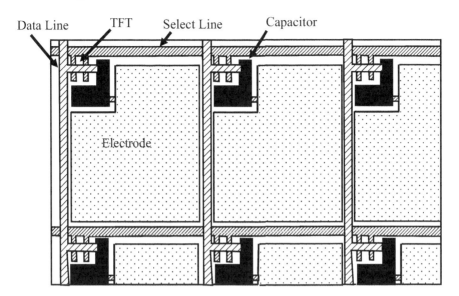

Fig. 5.2 Illustration of a portion of a substrate in an active matrix LCD

to be stored in a capacitor at each pixel. Data is loaded into the capacitor at each element by placing a signal on a select line attached to a row of TFTs in the display panel. The desired voltage is then provided on the data line. The capacitor at each row and column junction is charged to the voltage on the data line as the electric signal is transmitted through the TFT. The capacitor permits the voltage to be maintained at the electrode between updates of voltage at each row of circuits on the substrate.

Figure 5.2 illustrates a couple of important points. In this structure, the electrode is formed from a relatively transparent conductor, such as Indium Tin Oxide (ITO). Because this electrode is transparent, light can travel through this portion of the light modulator. The remaining components, including the data, select lines, and TFTs are typically formed, at least partially, from metals. Therefore, this portion of the light modulator does not transmit light. It is also important that based on the process used to form this substrate, the minimum dimensions of gaps and features within this structure are fixed. As a result, these gaps and features have a minimum size which might be nearly the same regardless of the resolution of the LCD. As the resolution of the LCD increases, the size of light emitting elements decreases. When the size of the light emitting element decreases, the relative area of the transparent electrode also decreases. As this is the only portion of the light modulator that transmits light, the area available for light to pass is decreased. Therefore the efficiency of the LCD decreases with increases in resolution. This is a particularly important point as the addition of differently colored light-emitting elements in a multi-primary display often requires the size of the light emitting elements to decrease, which effects the efficiency of the LCD.

> As the resolution of an LCD is increased, the manufacturing complexity generally increases and the power efficiency of the resulting display decreases.

Another important fact is that formation of the substrate shown in Fig. 5.2 requires the deposition of multiple thin film layers, as well as precision patterning of these thin film layers. Although not discussed, formation of this substrate requires insulating materials to be placed between the metal lines and deposition of a semiconductor, such as amorphous silicon (a-Si) or low temperature polysilicon (LTPS), between the layers in the TFT. As such, the formation of this substrate requires a significant amount of know-how. Further, manufacturing facilities capable of forming these substrates require significant capital investment. Because of the difficulty in forming these substrates, it is the ability to form these substrates which provide LCD manufacturing companies a competitive advantage in the marketplace. It is investment in these manufacturing facilities which have permitted companies such as Samsung and LG to dominate display manufacturing.

5.4 Contrast and Tone Scale

One of the more important characteristics of a display is the contrast of the display. Referring to the LCD structure shown in Fig. 5.1, it is important that most of the light from the backlight is blocked when light is linearly polarized in the first linear polarizer and then passes unimpeded to the second orthogonal linear polarizer. Unfortunately, some of the light entering the liquid crystal material is diffused or scattered, disrupting the polarization of this light. Additionally, neither of the polarizing films is ideal. Therefore, a small portion of the light passes through the polarizing films without having the required polarization. Therefore, a small portion of the light tends to pass through all light emitting elements in the LCD. As a result, even the light emitting elements intended to emit no light always emit at least a small amount of light. This then decreases the contrast of LCDs.

> As a small amount of light will always pass through even the black light emitting elements on an LCD, this light will limit the contrast that can be achieved from an LCD.

The tone scale of a typical desktop, active matrix, twisted-pneumatic LCD is shown in Fig. 5.3. Note that the minimum luminance that can be achieved with this display is 0.1884 cd/m^2. The maximum luminance of the display is 179.5 cd/m^2. As a result, the highest achievable contrast ratio is 953:1 corresponding to a contrast modulation of 0.998. As shown in Fig. 5.3 the tone scale of this monitor is compressive, with changes in low code values resulting in small changes in luminance while

5.4 Contrast and Tone Scale

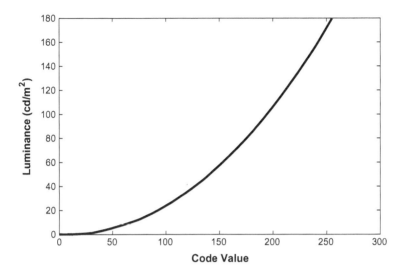

Fig. 5.3 A representation of the tone scale of a representative twisted-pneumatic desktop display

similar changes in high code values result in much larger changes in luminance. This relationship is not inherent in LCDs as changes in voltage result in a relatively linear change in light transmission in a typical twisted-pneumatic LCD. However, the mapping between code value and voltage is adjusted to provide the compressive relationship shown. This mapping is created to provide a display response which is similar to the sRGB tone scale, which was discussed in the previous chapter.

5.5 Viewing Angle

In the light modulator, the electric field arranges the liquid crystal molecules with respect to the substrate. However, not all light enters the light modulator at the same angle. Instead some rays enter the light modulator perpendicular to the substrate while other rays enter the light modulator at angles less than 90 degrees. Therefore, the direction of the light is different for light which enters the liquid crystal layer at different angles. Further, the thickness of the liquid crystal layer, while small, is large with respect to the wavelength of light. As a result, the path length of light which passes through the liquid crystal layer perpendicular to the substrate is shorter than the path length of the light which passes through the liquid crystal layer at smaller angles. As a result, light passing through the LCD perpendicular to the panel is modified differently than the light passing through the LCD at smaller angles. As a result, the light exiting the LCD changes as a function of viewing angle.

Figure 5.4 shows the tone scale and color variability of neutral measured perpendicular to a typical desktop twisted-pneumatic LCD as compared to the same display

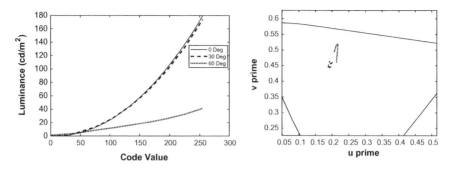

Fig. 5.4 Tone scale (left panel) and color (right panel) of neutral as a function of viewing angle

measured at angles of 30 and 60 degrees with respect to perpendicular. As shown, the black level, the white level, the tone scale and the color of neutral all change as a function of viewing angle for this display. Although the changes are relatively small as the viewing angle increases from 0 to 30 degrees, the change is dramatic as the viewing angle increases from 30 to 60 degrees. The luminance and luminance contrast of the display is reduced significantly for larger viewing angles. Further, the color variability as a function of tone scale becomes significantly larger. The amount of this viewing angle change differs across display technologies.

Table 5.1 shows the values for the twisted-pneumatic LCD as compared to a higher quality in-plane switching (IPS) liquid crystal display. As this table shows, both technologies are relatively stable between 0 and 30 degrees. The color variability is significantly less for the IPS display. However, the luminance contrast is reduced significantly as the viewing angle becomes large. For the twisted-pneumatic display, the contrast ratio decreases from 953:1 when viewed at zero degrees to only 35:1 when viewed from 60 degrees. This loss is slightly less for the IPS display, for which the contrast ratio decreases from 1446:1 to 319:1. While both of these displays are susceptible to loss in quality with increasing viewing angle, this loss is certainly less for the IPS display than for the twisted-pneumatic display. Other LCD technologies exist, which have degraded viewing angle performance as compared to these displays.

> Color and luminance from an LCD has traditionally varied significantly when one views a LCD from a different angle. Although this attribute of modern LCDs has improved dramatically, the available viewing angle is optimized for its application. For instance, an LCD designed to positioned in landscape mode and viewed as people walk by might have a wide horizontal viewing angle but might be designed to have a very limited viewing angle or will be nearly invisible if viewed from a certain angle (for example if viewed from below). As a result, various LCD models are produced for different applications providing broader horizontal or vertical viewing angles, depending upon their application.

Table 5.1 Table of black point luminance, white point luminance and white point uniform chromaticity coordinates for a representative twisted-pneumatic (TN) and in-plane switching (IPS) LCD panel at horizontal viewing angles of 0, 30 and 60 degrees with respect to a ray perpendicular to the display surface

Measurement	TN			IPS		
	0	30	60	0	30	60
Black luminance	0.1884	0.2835	1.18	0.1760	0.2256	0.1966
White Luminance	179.5	174.0	41.4	254.5	215.7	62.67
Contrast ratio	952.8	613.8	35.1	1446.0	956.1	318.8
Modulation	0.998	0.997	0.945	0.999	0.998	0.994
White u'	0.1888	0.2013	0.2067	0.2001	0.2004	0.2004
White v'	0.4542	0.4530	0.4617	0.4711	0.4728	0.4775

5.6 Response Time

Another important characteristic of LCDs is the time required for the liquid crystals to change alignment in response to a change in electric field. Traditionally, this has been an area of concern for LCDs. Significant improvements have been made for this attribute of LCDs with most modern displays advertising response times of less than 8 ms and many displays advertising response times of 4 ms or less. However, this attribute can provide noticeable artifacts under some circumstances. For instance as the temperature of the display is reduced, the mobility of the liquid crystals can be reduced. Under these circumstances the switching time of the LCD can increase, resulting in visible delays in image update. Response time for some LCDs continue to introduce blur for objects which move rapidly across a display, even when viewed in environments with controlled temperatures.

5.7 LCD Innovations

Having discussed the structure and performance characteristics of LCDs, it is also useful to discuss areas of innovation for LCDs. Specifically we will discuss innovations occurring in the LCD backlight and drive mechanisms to improve display performance. As we will see these innovations are primarily targeted at improving display saturation, decreasing power consumption, and increasing perceived contrast of the display.

5.7.1 Backlight Innovations

One of the primary areas of an LCD which is currently undergoing innovation is the backlight and accompanying LCD drive techniques. As we have discussed, the backlight has traditionally been formed from fluorescent lamps. In recent years, these backlights have been replaced with white LEDs. As discussed in Chap. 3, the iLEDs used in LCD backlights are generally formed from a blue iLED which is coated with a yellow phosphor to produce a white light backlight.

Referring to Fig. 3.11 one can see that these white LEDs emit light having a distinct blue peak but broad emission across longer wavelengths. This characteristic of white LED emission is desirable when these LEDs are applied in a general purpose light source as light is emitted at most visible wavelengths permitting objects in the environment to reflect light at almost any visible wavelength. However, this characteristic of white LEDs is not desirable when LEDs are applied in a display. Instead of desiring broadband emission, in a RGB display it is desirable to create saturated red, green, and blue light emitting elements. As we discussed in Chap. 3, to obtain saturated color, the green color filter must be narrow and the red color filter must emit a subset of the long wavelength light. For example, a typical LCD might employ red, green, and blue color filters having spectra such as shown in Fig. 5.5. As noted, these spectra filter the majority of the energy from the light source.

Table 5.2 shows the uniform chromaticity coordinates, saturation, and the proportion of energy passed for each of the red, green, and blue light emitting elements. As shown, the color filters over the red and blue elements each emit about $1/4^{th}$ of the light which reaches the color filters. About $1/3^{rd}$ of the light is passed by the green

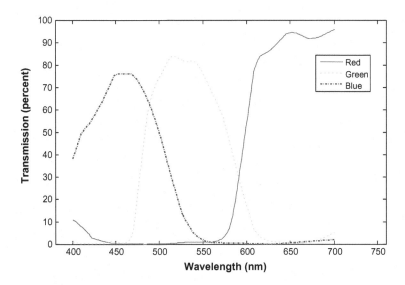

Fig. 5.5 Example LCD color filter spectra for red, green and blue elements [3]

Table 5.2 Parameters for red, green, and blue filtered white LEDs. Uniform chromaticity coordinates; saturation and efficiency are shown for each primary color assuming a backlight having the "Cool White" spectra shown in Fig. 3.11 and color filters having spectra shown in Fig. 5.5

Primary	u'	v'	Saturation	Relative efficiency (%)
Red	0.443	0.530	3.27	21.3
Green	0.127	0.565	1.59	33.3
Blue	0.154	0.202	3.49	28.2

color filter, however the saturation of this color is significantly lower than the saturation for the red and blue light emitting elements. This fact illustrates a significant tradeoff when using color filters to form red, green, and blue primaries using color filters. That is, there is a significant tradeoff between the efficiency of the light passed through the filters and the color saturation of the primaries. This trade is particularly harsh when the primaries are formed by filtering broadband white light, as is the case in this example. That is, increasing the saturation of the primaries by selecting narrowband color filters imposes a significant cost in terms of power efficiency of the individual primaries. As the display has only red, green, and blue primaries, reducing the power efficiency of the primaries translates to a loss in power efficiency of the display.

> When applying color filters to form RGB displays, the color filters typically absorb at least 2/3rds of the energy available in an underlying white light source, permitting less than 1/3rd of the light to pass to the observer. As the color filters are redesigned to create greater saturation, they absorb even more light from the white light source.

5.7.2 Quantum Dot LCD

As we discussed in Chap. 3, we can employ other color change materials in the backlight, including quantum dots, which have narrow light emission bands. Lights with three narrow peaks as shown in Fig. 3.12 is not generally desirable for general lighting. These lights have significant bands within the visible spectrum, where little, if any, light is emitted, causing reflective objects having certain colors to appear darker than they should. However, lights with narrow emission spectra can be desirable in display backlights. Table 5.3 shows similar values as Table 5.2 with the exception that the backlight uses a white based upon a blue LED and red and green quantum dots as shown in Fig. 3.12. Note that the use of quantum dots improves the saturation without having a significant effect on the relative efficiency of the primary colors, as

Table 5.3 Parameters for red, green, and blue filtered white LEDs. Uniform chromaticity coordinates; saturation and efficiency are shown for each primary color assuming a backlight having the spectra shown in Fig. 3.12 and color filters having spectra shown in Fig. 5.5

Primary	u'	v'	Saturation	Relative efficiency (%)
Red	0.510	0.517	4.32	26.1
Green	0.077	0.577	2.48	27.2
Blue	0.191	0.424	3.34	30.9

compared to the white emitter depicted in Table 5.2. These calculations thus illustrate that a backlight having the peaked, narrow-band emission typical of the quantum dot backlight can be more easily filtered to maintain the majority of the light from each of the three peaks without losing significantly more than two thirds of the energy to the color filters. Notice that even for the best backlight and color filter combination, if our goal is to obtain saturated red, green, and blue light from a white backlight, the color filters will need to eliminate the energy from two of these color bands, passing one third or less of the light. Therefore, a display using color filters to form colored light emitting elements from a white backlight can never provide saturated light without color filters that absorb $2/3^{rds}$ of the light which impinges upon them. A comparison of Tables 5.2 and 5.3 illustrates the potential for quantum dot or other narrow emission band backlights to improve color saturation and simultaneously improve power efficiency of LCDs.

> In existing displays the quantum dot layer is placed behind the LCD as part of the backlight. Recent announcements from quantum dot manufacturers have discussed the use of a patterned quantum dot layer to replace the color filters in an LCD having a blue backlight. Such a configuration will significantly increase the efficiency of future LCDs but will require the use of a yellow color filter in conjunction with the quantum dots to reduce blue leakage through the red and green light emitting elements, as well as preventing ambient light from exciting the red and green quantum dots, which would reduce display contrast.

5.7.3 HDR LCD

The use of color conversion materials with sharp spectral peaks, such as those provided by quantum dots described in the previous section, enhances the color saturation of traditional LCDs. These LCDs still suffer from less than desired contrast as unwanted light leaks through the light emitting elements that are intended to block light traveling from the backlight to the user. It is important, however, that these back-

5.7 LCD Innovations

lights can include two-dimensional arrays of iLEDs rather than fluorescent lamps. Remembering some of the characteristics of these technologies, an important property of iLEDs is their ability to be turned on and off without noticeable delay. This differs from fluorescent lamps which often require a period of time to be turned on to full luminance. As a result, it is possible to adjust the luminance output of each iLED instantaneously or to even turn each iLED on or off behind the LCD. Because these iLEDs are arranged in a two-dimensional array and each iLED provides light to only a small area on the LCD, the current to these iLEDs can be manipulated to control the amount of light the backlight creates at various areas behind the LCD.

The ability to dim areas of a low resolution light source behind an LCD and still create a high quality image was demonstrated in the early 2000s [10]. Originally this concept was envisioned as a method for dramatically increasing both the contrast and luminance of an LCD without significantly increasing the power consumption of the display. In this concept, iLEDs in dark or black areas of an image were dimmed or turned off while iLEDs in bright areas of an image were switched to higher luminance than used in a typical LCD. Later this same concept was employed to simultaneously increase the contrast of an LCD while reducing the power consumption of an LCD. In this implementation, iLEDs behind dark image areas are dimmed, while iLEDs behind bright image areas were switched to a similar luminance as a typical LCD. However, because image content has an average of around 20% of white, most of the iLEDs in the backlight are dimmed compared to their output in a traditional LCD backlight. In this condition, the amount of unwanted light emission is reduced because the backlight is dimmed or even turned off in dark areas of an image, producing less light to leak through the LCD layer.

To illustrate this concept, we will start with a band from an image as shown in the top band of Fig. 5.6. We then assume that an array of 4 by 16 iLEDs lie behind this image band. To simplify the problem, we will assume that each iLED creates a uniform luminance within each block. It should be recognized, however, that each iLED likely produces a distribution of light approximating a two-dimensional Gaussian. In this example, one can imagine then selecting the brightest light emitting element within each of 10 blocks across the band in Fig. 5.6 and adjusting the luminance of each of the iLEDs to produce the maximum luminance within each block. This might produce a backlight having a pattern such as shown in the second band of Fig. 5.6. Finally, the LCD would be controlled to create the correct image on the LCD. The LCD might appear as shown in the bottom band of Fig. 5.6. When the LCD shown in the bottom band is illuminated by the backlight shown in the 2nd band, the image at the top is produced. Note that the middle band is not completely white but instead contains a number of gray or black blocks, reducing the power consumption required of the backlight and increasing the perceived contrast of the final display.

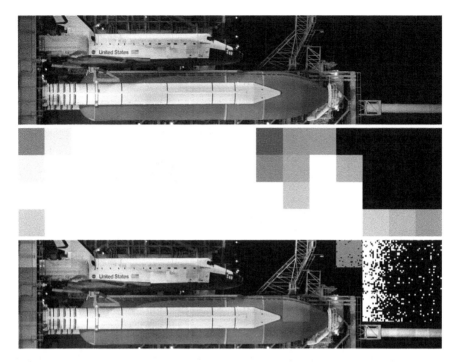

Fig. 5.6 Illustration of a portion of an image shown in the top panel, a backlight pattern used in an HDR LCD shown in the middle panel, and a LCD pattern useful for creating the image with the backlight, shown in the bottom panel

Modulating both the backlight and the LCD in real time can produce LCD displays with higher perceived contrast and lower power consumption. However, without appropriate design, unwanted image artifacts can be introduced and the contrast in small areas of the display is physically less than desired, although this loss of contrast can be masked by flare in the human eye to the extent that it may not be noticeable under most circumstances.

5.7.4 Temporally Modulated LCD

A significant advantage of LED technology is the ability to instantaneously switch the LED from a light-emitting state to a dark state and vice versa. As a result, it is possible to use differently colored iLEDs as a light source for a LCD and sequentially illuminate the LCD with red, green, and blue light. If the LCD is switched synchronously with the iLEDs so that each light-emitting element within the LCD

5.7 LCD Innovations

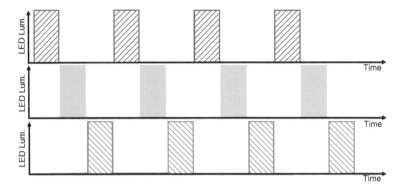

Fig. 5.7 Light output of red, green, and blue LEDs as a function of time within the LCD light source

is switched to provide the red, green, or blue portion of the image at the same time the appropriately colored iLED is on, a full color image can be provided without the need for color filters. This method clearly has the advantage that the color filters discussed earlier are no longer necessary. As such, the display can be much more energy efficient and as the iLEDs emit light with a narrow spectral emission, the display can provide relatively pure color. The elimination of the color filters reduces the manufacturing cost of the LCD, in addition to significantly increasing the power efficiency of the display.

Figure 5.7 illustrates the time sequence with which the red, green, and blue LEDs within an LCD light source would need to be illuminated to facilitate such a display. As shown, the red, green, and blue LEDs are each turned on for a short period of time. Further, when one color of LED is activated, the remaining colors of LEDs are deactivated. Usually, there is also a gap between the times that one colored iLED is activated and the times that the next color of iLED is activated to provide the LCD some time to switch. These displays follow this time-sequenced pattern where each color is activated before the next color and the sequence is repeated many times per second. Because each color field (red, green, or blue) are each sequentially presented, this class of displays are commonly referred to as field-sequential displays.

Unfortunately, this approach is not without its problems. As we mentioned earlier, the switching time of most liquid crystal materials is relatively slow. As such, the switching time of the liquid crystal material limits the number of times per second that the color can be switched from red, to green, to blue. Certain types of LCDs, do permit higher switching rates. However, these devices also tend to have a relatively narrow viewing angle. These LCDs can be backlit, but reflective devices are also available. Further, the faster switching LCDs often tend to be manufactured on relatively small substrates. As such, they often compete with microelectromechanical displays, such as the digital micro-mirror devices [5]. Therefore, they are often implemented within projectors where the viewing angle of the device is not important.

Generally direct view field sequential LCDs have not been practical due to the switching time of the LCDs. If the color is switched too slowly, the user can see flicker. While flicker may not be present if the display is updated at a rate of 60 Hz per color field (180 fields per second for a three color RGB display), an image artifact referred to as color breakup is often quite visible in these displays. This artifact occurs either as fast moving objects are presented on the display [6] or a user moves their eyes across the display [8], causing different colors of the same object to be presented to different locations of the human retina. Therefore, the human sees flickering or blurred color along the edges of objects, an artifact which can be quite disturbing. Research has suggested that field rates as high as 480 Hz are likely necessary to eliminate this artifact within traditional three-color displays [9]. Therefore, three color field-sequential color LCDs are generally not practical, despite the power consumption advantages provided by this class of display.

5.8 Summary and Questions for Thought

In this Chapter, we discussed the origin of the LCD. As part of this discussion we introduced the general structure of the LCD. Throughout this discussion we reviewed the fact that many components within a typical LCD are manufactured by different companies, providing significant innovation but distributing the profits associated with LCD manufacturing across various entities. We also discussed the importance of the electronics constructed to drive the LCD and the implications these electronics have on power consumption of the LCD as the resolution of the display changes. Additionally, we discussed some of the important characteristics of typical LCD technologies, including their contrast, viewing angle, saturation, and temporal characteristics. Although we did not discuss the history of these display characteristics, it is these characteristics, as well as the resolution of LCDs which have improved dramatically with innovation in LCDs. Finally, we discussed some recent innovations in backlight technology which have provided significant advances in LCD image quality.

It is important as we leave this section to recognize the LCD is the dominant display technology within today's display market. For competitive technologies to displace this technology, it is necessary for these competitive technologies to have a significant competitive advantage over LCDs. This advantage is likely to include image quality, power consumption, or manufacturing cost. However, without a significant advantage along one or more of these attributes, any competing technology is likely to face overwhelming competition from the dominant technology.

Looking back at this chapter, a few questions for thought can be proposed:

(1) When is it useful to have a large number of companies contributing to the development of underlying loosely integrated technologies verses fewer companies contributing to a highly integrated technology?

(2) What are the primary vulnerabilities of LCD technology which may make it possible to replace this technology?
(3) Assuming that multi-primary LCDs can be readily produced which have four or more colors of light-emitting elements, given the fact that adding light-emitting elements reduces the aperture ratio of the light-emitting elements, reducing their efficiency, when is it really practical to include additional light-emitting elements? Is there any way to overcome the reduction in aperture ratio?
(4) Can the use of multi-primary displays impact the utility of advanced display concepts, such as field sequential LCDs?
(5) Longer term, what innovations are likely which will provide higher efficiency LCDs?

References

1. Dunmur D, Sluckin T (2011) Soap, science, and flat-screen TVs: a history of liquid crystals. Oxford University Press, New York, NY
2. Edwin HL (1934) Light valve and method of operation. United States Patent Number 1,955,923
3. Hamer JW, Arnold AD, Boroson ML, Helber MJ, Levey CI, Ludwicki JE et al. (2008) System design for a wide-color-gamut TV-sized AMOLED display. J Soc Inf Disp 16(1):3–14. http://doi.org/10.1889/1.2835033
4. Helfrich W, Schadt M (1970) Light control cell. Swiss Confederation Patent Number 532,261
5. Hornbeck LJ (1991) Spatial light modulator and method. United States Patent Number 5,061,049
6. Johnson PV, Kim J, Banks MS (2014) The visibility of color breakup and a means to reduce it. J Vis 14(2014):1–13. https://doi.org/10.1167/14.14.10.doi
7. Lueder E (2001) Liquid crystal displays: addressing schemes and electro-optical effects. Wiley, Chichester
8. Mori M, Hatada T, Ishikawa K, Saishouji T, Wada O, Nakamura J, Terishima N (1999) Mechanism of color breakup in field-sequential-color projectors. J Soc Inf Disp 7(4):257–259
9. Post DL, Monnier P, Calhoun C (1997) Predicting color breakup on field-sequential displays. In: Proceedings of the SPIE 2058, head-mounted displays II
10. Seetzen H, Whitehead LA, Ward G (2003) A high dynamic range display using low and high resolution modulators. In: SID symposium digest of technical papers, vol 34, p 1450. http://doi.org/10.1889/1.1832558

Chapter 6
LED Display Technologies

6.1 Organic Light-Emitting Diode (OLED) Displays

The first low power Organic Light-Emitting Diodes (OLEDs) were discussed publicly by Ching Tang and Steve Van Slyke in 1987 [16]. An OLED creates light when an electric field is created across layers of a few hundred angstroms of organic chemicals coated between a pair of electrodes. Although emission from organic materials was known, this emission was very low in efficiency until a pair of organic materials were coated between electrodes. In this configuration one of the materials promoted the flow of electrons away from the cathode and the other organic material promoted the flow of flow of electrons towards the anode, producing holes for the electrons from the cathode to fill. Efficient light emission then occurred at the boundary between the two organic materials.

Passive matrix displays based on Organic Light-Emitting Diodes were first introduced as very low resolution character displays in the mid-1980s. These displays were capable of providing very sharp, bright, high contrast characters through a very thin, light-weight emissive display format. Unfortunately, the technology was limited to very low resolution and competed directly with reflective LCD displays, which required very little power to operate. Therefore, the volume of these displays demanded by the market was limited. Their primary advantage was the fact that they emitted bright, high contrast light.

During the mid-1990s, full color imaging display prototypes were developed based upon active-matrix OLED technology by numerous companies, with the first commercial active matrix display being produced in a joint venture between Eastman Kodak Company and Sanyo Electronics. By 2002, the first desktop display prototypes were demonstrated by Eastman Kodak Company and Sanyo Electronics. The first television-sized (i.e., 40 in. diagonal) OLED display was demonstrated by Samsung in 2004.

Compared to LCD, the number of components to be manufactured in an OLED display are relatively small. OLEDs consist of organic materials, deposited on a backplane, often covered by an optical film. These displays may be encapsulated and often will contain some type of desiccant to keep moisture from entering the display. This is important as many of the materials within the OLED display are highly reactive with oxygen. Therefore, companies contributing to this industry include a few companies producing organic materials, companies capable of producing high quality substrates (typically LCD panel manufacturers), and optical film producers to provide the single optical film on the front of the OLED. Some OLEDs can also include color filters, requiring color filters similar to those produced for LCDs. As such the number of components to be manufactured for this display are significantly fewer than required for an LCD. Further, the value to be extracted from OLED manufacture is heavily controlled by the companies capable of providing high quality backplanes and somewhat by companies capable of providing high quality organic materials. However, the precision required to manufacture a high quality OLED display can be significantly higher than that required to form an LCD and, therefore, adoption of this technology has been limited predominantly by the availability of robust manufacturing techniques.

> OLED displays are highly integrated, solid state devices. Unfortunately, the manufacturing complexity of an OLED display is significantly higher than it is for a typical LCD.

6.1.1 Technology Overview

By its broadest definition, an OLED is a diode that includes two-or-more, thin-film, organic layers, including at least one light-emitting layer. To form an active diode, the thin films of organic materials are deposited between a pair of electrodes as shown in Fig. 6.1. The electrodes include a cathode for injecting electrons (e-) into the organic layers and an anode for injecting holes (h+) into the organic layer. Light is formed when the electrons and holes recombine in the organic light-emitting layer, creating an excited state on an organic molecule, which then relaxes to its ground state, providing a photon during relaxation. In most OLEDs, at least one of the electrodes is formed from a metal oxide, such as Indium Tin Oxide (ITO) or a very thin metal, to create a transparent or semi-transparent electrode. The remaining electrode is formed from a reflective metal, such as aluminum or silver. As shown in Fig. 6.1, this arrangement permits light, which is emitted within the OLED according to a Lambertian distribution, to be channeled to one side of the diode to be viewed by the user. The OLED structure itself is typically less than 2000 Angstroms in thickness, about $1/100^{th}$ the thickness of a human hair.

To efficiently create light, the common OLED includes additional electron transport and hole transport materials as shown in Fig. 6.1, which permit the electrons and

6.1 Organic Light-Emitting Diode (OLED) Displays

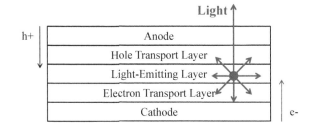

Fig. 6.1 General OLED structure, layer thicknesses are not to scale. Figure illustrates layer order and arrangement

holes to migrate away from the electrodes to the light-emitting layer located between the two layers of transport materials. The electron transport layer is important to prevent the energy from being channeled along the metal electrode surface and lost through surface plasmon effects. Other material layers are also important to reduce energy level gaps between the electrodes and the light-emitting layer and to reduce the resistance of the OLED, lowering the electrical potential necessary to support light emission. Further, the light-emitting layer is typically formed from multiple materials, to include host and dopant materials, which improve energy efficiency and lifetime of the resulting organic light-emitting diodes.

During manufacturing, the organic layers are typically formed in the device through vapor deposition. In this process the materials and the display substrates are placed in a vacuum. The small molecule materials are then heated to a temperature above their sublimation temperature permitting the solid organic materials to be vaporized. The resulting vapor then rises before condensing on the cooler display substrate. This process requires specialized manufacturing equipment but permits the formation of multiple, specially designed layers, which individually support functions. Desired functions include electron injection, electron transport, hole blocking, light-emission, and hole transport among others. As such, performance of the final device can be highly optimized through the use of several independently-deposited layers. Further, it is possible to blend materials between material interfaces. Because this manufacturing process includes vapor deposition, this deposition can be performed across very large areas, permitting the production of large displays.

Alternate manufacturing processes include printing. Early printing required the use of polymer organic materials [2]. This process has the potential to be less expensive as it does not require the formation of a vacuum or precise temperature control. Materials can be applied with precision to desired locations rather than being released in a plume, which coats not only the entire substrate but the inside of the vacuum chamber as well. However, each layer that is added during a printing process has the potential to dissolve any existing layer, which restricts the number of organic layers that can be formed within a printed OLED. Small molecule devices typically outperform polymeric devices along one or more criteria, which can be at least partially attributed to the constraints of the manufacturing process. In fact, modern printing approaches for OLED include small molecule organic materials. However, the resulting devices often have lower power efficiency and shorter lifetime than current vacuum-deposited small molecule devices.

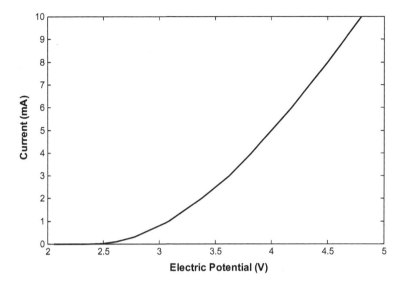

Fig. 6.2 Current as a function of electrical potential for a typical OLED. Data based on [14]

High quality OLEDs typically include multiple layers of organic material with each layer having a specialized function. While vapor deposition permits the application of several organic layers, the number of layers are often limited by other manufacturing processes, including printing.

6.1.2 Electric Properties

For the OLED to generate light it is necessary to form an electric field across the organic layers and for current to flow through the organic materials. As the electric potential across the OLED is increased, a threshold voltage will be attained, above which current begins to flow through the device, permitting the production of light. Figure 6.2 depicts the flow of current through a typical red-light emitting, small molecule florescent OLED as a function of the electrical potential between the anode and cathode [14]. As this figure shows, an electrical potential just over 2 V is required before current begins to flow through the device. Above this voltage, the flow of current increases nonlinearly with further increases in voltage. The voltage response of a typical LED can be characterized as shown in Eq. 6.1.

$$I = I_o \left(e^{(V/k)} - 1 \right) \tag{6.1}$$

6.1 Organic Light-Emitting Diode (OLED) Displays

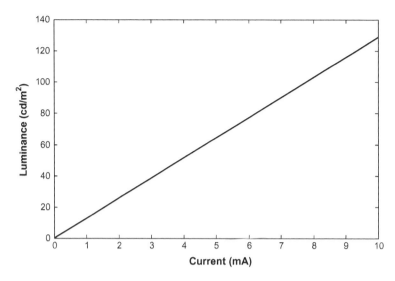

Fig. 6.3 Luminance as a function of current for a typical OLED

In this Eq. 6.1 represents current, I_o represents the initial current, V represents voltage and k is the fitting constant. The general formula shown provides a limited fit to the data shown in the following figure, due primarily to the absence of an intercept. Adding the intercept for current values greater than zero provides a better fit. The equation used to fit the data in Fig. 6.2 is shown in Eq. 6.2.

$$I = 8.641\left(e^{(V/1.594)} - 5.121\right) \tag{6.2}$$

A general function of this form fits many typical OLED current-voltage (IV) functions.

Since light is produced as a function of the number of electron-hole pairs created in the light-emitting layer, the luminance output of the OLED is linearly-related to current with the slope of the line representing the efficiency of OLED device. A common relationship between current and luminance of a device is as shown in Fig. 6.3. As shown, the function is linear with an intercept near zero and a slope indicative of efficiency. For example, the efficiency of the device depicted in Fig. 6.3 is 12.9 cd/mA.

Despite the differences in efficiency and threshold voltage, the shape of the curve relating the electrical potential across the diode to the flow of current through the diode is generally consistent across colors and types of OLED devices. As a result, this nonlinear current to voltage relationship and the linear luminance response to current relationship produce a relationship between luminance and electrical potential across the diode which follows a nonlinear relationship that is similar to the current to voltage relationship. For example the same device depicted in Fig. 6.2 will have a luminance

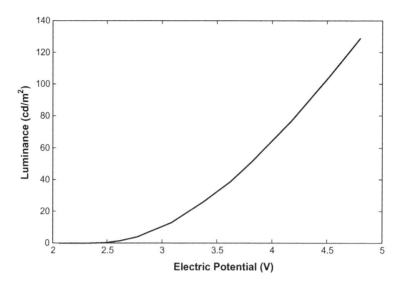

Fig. 6.4 Luminance as a function of electrical potential for a typical OLED

to voltage relationship as depicted in Fig. 6.4. Once again, this nonlinear response is only present as a function of electric potential across the diode.

The electrical performance of OLEDs has significant implications for perceived bit depth (i.e., the ability to provide the appearance of a continuous tone image with a minimum number of physical bits), in electronic displays. The human visual system has a nonlinear response to light that is often modeled as logarithmic or power functions, as discussed in Chap. 2. As such, the human observer is able to distinguish small changes in relative luminance levels near black but only relatively large changes in illumination at high relative luminance levels. As a result, to achieve a display with high perceived bit depth it is desirable for the display to present information such that luminance is nonlinear with respect to the control signal. The luminance of an OLED can be modulated by controlling either the proportion of time the OLED emits light using a pulse width modulation (PWM) scheme, the current provided to the OLED using a current control scheme or by controlling electric potential using a voltage control scheme.

If an OLED display is controlled using either PWM or current drive, the luminance response of the display will be linearly related to the drive signal. Due to the linear response of the display, each increase in bit value will correspond to an equal change in display luminance. Therefore, a large number of bits will be required to provide the perception of high bit depth at low luminance level, even though changes in bit value will result in perceptually unnoticeable changes in luminance in the high luminance regions of a displayed image. The nonlinear response of the OLED in the voltage domain makes this domain much more favorable for attaining displays with a high perceived bit depth because the nonlinear response of the OLED is complimentary to the nonlinear response of the human visual system. Specifically assuming that it

6.1 Organic Light-Emitting Diode (OLED) Displays

is necessary to control the display to provide differences in perceived lightness equal to 0.5 L* values throughout the entire tone scale, 12 bits would be required to drive a display having a dynamic range from 0.1 to 500 cd/m^2 with a linear response while 9 bits are required for a voltage driven display.

> Luminance output from an OLED is approximately linear due to the current flow through the OLED while luminance is a power function of the voltage across the OLED.

6.1.3 Color Performance

Color formation in an OLED is achieved primarily through molecule selection. Molecules exist which emit light across the visible spectrum and even into the near infrared portions of the electromagnetic spectrum. Therefore, to create a full-color display, it is necessary to pattern at least one of the electrodes (e.g., the anode shown in Fig. 6.5) and the light-emitting layer. As a result, a display structure for a bottom-emitting display (e.g., a display that emits light through the substrate) might appear as shown in Fig. 6.5. As will be discussed in more detail later, this structure can present manufacturing issues as it is necessary to pattern the red, green, and blue emitters in a side by side arrangement in alignment with the anodes. In existing manufacturing, this is accomplished by depositing materials through a metal shadow mask. The shadow mask is typically a metal plate having holes machined through to permit light-emitting material designed to emit one color of light to pass through it to be deposited on the desired regions of the substrate. These shadow masks are subject to thermal expansion and contraction during the deposition process as it is placed between the organic materials which are being heated to a temperature above their sublimation temperature and the substrate which must be cool enough to permit the organic materials to condense on the surface of the substrate. Precise alignment of the shadow mask to the substrate is difficult, especially as the shadow mask expands and contracts during deposition. The inability to maintain precise alignment creates defects. As a result, the resolution of displays that can be achieved with this process can be limited. Further, the size of the substrates or shadow masks used in manufacturing can be limited as well. This is important as the LCD industry has reduced material handling cost in display manufacturing through the use of very large substrates and to be cost competitive, it would be helpful if OLEDs could be manufactured on very large substrates as well.

In addition to manufacturing limitations, OLEDs can suffer color purity issues as well. Vacuum-deposited OLEDs are formed from an unstructured, loosely coupled matrix of molecules connected only through Van der Waals forces. This process can theoretically be much less expensive than the formation of iLEDs into a structured

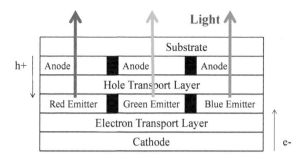

Fig. 6.5 Depiction of a full-color light emitting element in an RGB patterned display. Note layers are not shown to scale. Figure illustrates the general arrangement of patterned anodes and emitting layers

matrix. Unfortunately, the lack of structure comes with a cost. While iLEDs emit primarily at a single frequency with a low amount of variability, permitting a spectral peak with a Full Width at Half Maximum amplitude (FWHM) of about 30 nm or less, OLED emitters can, and often do, emit light at multiple frequencies or have significantly wider spectral peaks. Figure 6.6 shows the spectra for three OLED emitters that have useful color coordinates. As shown, the full width at half maximum amplitude (FWHM) of the emitter peaks are approximately 100 nm for blue, 70 nm for green and about 45 nm for red. Each of these emission spectra is broader than typical inorganic LEDs. Further, while the iLEDs typically have a single mode of light output, resulting in a single peak in their emission spectra, the emission spectra for OLEDs often contain multiple emission peaks, which can further degrade the resulting color quality. Finally, the spectra of OLED emitters are often not symmetric, exhibiting longer tails on the long wavelength side of the spectral distribution.

Converting the spectra of these three emitters to uniform chromaticity coordinates, these values can be compared against the standard aim chromaticity coordinates as discussed in Chap. 4. This comparison is made in Fig. 6.7 with the broken lines indicating the gamut specified by the standards and the solid line triangle showing the gamut of a display employing the three emitters whose spectra are shown in Fig. 6.6. As shown, the blue and green coordinates for the three OLED primaries do not fall near the aim color coordinates. Further, the red primary for the OLED is inside the gamut specified by the DCI-P3 standard. These emitters do not represent the best of the OLED emitters for displays as these emitters were originally designed to be used in combination with color filters to form a display. However, these example emitter spectra indicate a potential issue with OLEDs. Specifically, the OLED emitters tend to provide broadband emission and often fall short of meeting the most desirable chromaticity coordinates within color display standards.

As a result, during the early evolution of OLED displays, it was difficult to construct an OLED display with a high color gamut. However, progress has been made in this area in recent years. For example, research in fluorescent red OLED devices has yielded red light-emitting devices having peak wavelengths at 608–609 nm

6.1 Organic Light-Emitting Diode (OLED) Displays

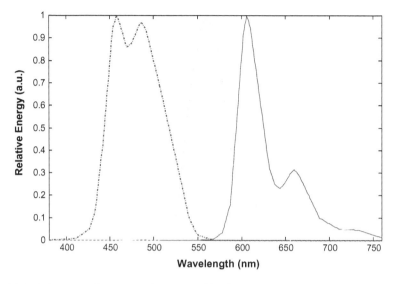

Fig. 6.6 Relative energy for typical red, green, and blue florescent OLED emitters. Emission spectra based on data from Murano and colleagues [12]

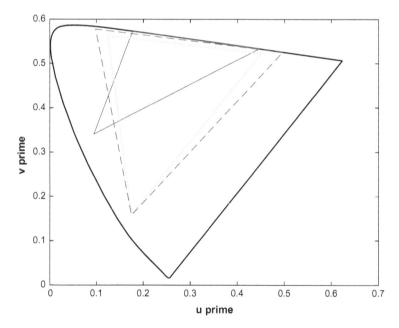

Fig. 6.7 Uniform chromaticity diagram showing chromaticity coordinates for an older OLED material spectra as compared to reference standards. OLED example gamut is shown by the solid red line, REC 709 is shown in the bold dotted green line and P3 by the dashed blue line

with FWHM bandwidths of about 25 nm providing narrow bandwidth emitters [8]. Although the spectral shape of the emitters is not publicly available, assuming that the emission follows a Gaussian shape, the chromaticity coordinates for these devices can be calculated as 0.442, 0.534. While this paper demonstrates the ability to narrow the spectrum of red emission, the relatively short peak wavelength of 609 nm prevents this primary from expanding the gamut as compared to the standard, two-peaked red discussed earlier. Similar devices have been demonstrated with narrow emission and purer color coordinates. For example, Kwong and colleagues demonstrate red emitters with uniform chromaticity coordinates of 0.527, 0.520 [9]. In recent years, similar advances have been demonstrated for blue and green emitters, providing gamuts near or exceeding the SEMPTE C standard.

While OLED materials traditionally have a broader spectrum than iLEDs, leading to displays with reduced color saturation, recent improvements in organic materials and the inclusion of optical effects through microcavity structures have permitted significant improvement in OLED color saturation without loss of efficiency.

This discussion of color, however, only pertains to OLEDs which are fabricated with at least one transparent electrode. When discussing color performance of an OLED, it is important that the entire thickness of an OLED emitter is often on the same order as the wavelength of visible light. At this thickness, the organic materials are relatively transparent. As such, it is possible to select material thicknesses and electrode properties to form OLED emission within a resonant Fabry-Perot microcavity.

In a microcavity, when the wavelength of the light matches the thickness of the optical cavity, a standing wave is produced and the light provided by the emitting layer is altered through constructive and destructive interference to produce a preferred wavelength of light. In OLED microcavities, the properties of the user-facing electrode can be modified to increase or decrease its reflectivity, effectively increasing or decreasing the strength of the interference effects. In a typical OLED, which includes one reflective electrode and a second electrode formed from a relatively transparent material, such as ITO, any microcavity effect will be quite weak. It is clear that even in these devices certain wavelengths of light are preferentially emitted. However, if some or all of the ITO is replaced with a reflective metal, for example silver, the microcavity effect becomes stronger as the reflectivity of this electrode is increased.

Figure 6.8 illustrates the general structure of a full-color display exhibiting a microcavity. Note the thickness variation in the electron transport layer (ETL) used in this example will impart a variation in the thickness in the material layers to create microcavities having different preferred wavelengths of emission. Note that it would typically be necessary to provide different thicknesses to each of the three emitter structures. However, some implementations have provided a red microcavity with a mode 1 or 2 cavity (employing a cavity that is one or two times the wavelength of the red emitted light) but provided the blue emitting layer with a mode 2 or 4 microcavity (a cavity that is 2 or 4 times the wavelength of the light). Using this trick, the thickness of these two cavities can be tuned to be equal. Nevertheless, at least the green cavity will require a different thickness, making it necessary to pattern

6.1 Organic Light-Emitting Diode (OLED) Displays

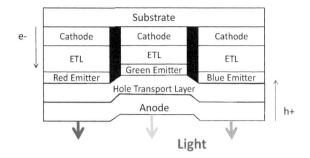

Fig. 6.8 Illustration of the layer structure for a top-emitting patterned RGB display, having a microcavity. Note layer thicknesses are not to scale but illustrate patterning and thickness variations

at least one layer to modify the thickness of this layer. Also note that microcavities are typically top-emitting, with the layers flipped and the electronics on the substrate controlling the flow of electrons. The anode is then typically formed from a bilayer of ITO and a reflective metal to form the resonant microcavity. Although the figure shows the thickness of the ETL changing, other layer thicknesses can be changed. For example, ITO could be patterned over the electrode on the substrate (cathode in Fig. 6.8) to impart a variation in cavity thickness.

The presence of this cavity has multiple effects. First the emission spectrum of the light emitted by the OLED is narrowed, which can improve the color purity of the emitted light. This is illustrated in Fig. 6.9, which shows the spectrum of a typical green emitting material formed in a bottom-emitting configuration with ITO as the anode which produces little microcavity effect and the same material formed in a top-emitting configuration with a metal anode and partially reflecting cathode including a Ag/Mg layer, which produces a stronger microcavity effect [13]. As shown, the microcavity structure significantly narrows the spectra of the emitted light, reducing the FWHM of emission from 74 to 46 nm. While this effect permits the purity of the light emitted to be enhanced, to acquire efficient emission the peak wavelength of the microcavity must correspond to the peak in the emission spectrum. This requirement limits the ability to use this effect to improve the color gamut of the display. However, by narrowing the distribution of light, the color gamut of the display can be improved significantly. For example, the device depicted in Fig. 6.9 is improved from uniform chromaticity coordinate values of (0.139, 0.568) to (0.104, 0.577), resulting in a significantly more saturated green. Similar effects can be obtained for blue and red devices through the use of a microcavity.

A second advantage of a microcavity is that an optical gain is typically observed, at least in the axis perpendicular to the face of the microcavity. This optical gain can improve the on-axis efficiency of the OLED by a factor of 2 or greater for strong microcavities. For instance, the example devices depicted above have efficiencies of 71 cd/A for the bottom-emitting device and 140 cd/A for the top-emitting (microcavity) device. Each of these attributes can be highly desirable when applying OLEDs to form a visual display. Unfortunately, the formation of microcavities creates several challenges when producing a visual display.

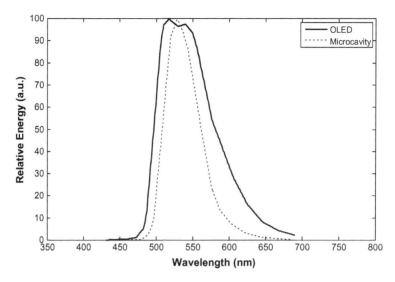

Fig. 6.9 Example spectra of a bottom emitting OLED with negligible microcavity and a top-emitting OLED including a microcavity

In displays using microcavities, the OLED device thickness must be color dependent, with each different color of emitter in a display requiring either a different OLED thickness or requiring an inorganic spacer to be formed in each emitter. This requirement adds additional manufacturing steps. Secondly, manufacturing tolerances for OLED device thickness variation are reduced significantly. Because the peak wavelength of the emitted light is selected by the thickness of the device, a relatively small unintended change in the thickness of the OLED across the display substrate will shift the resonant peak wavelength and produce color variation across the substrate. Third, additional constraints are placed on the electrode materials as there is a need for both electrodes to include a reflective material. For example, silver is significantly higher in reflectivity than aluminum and will produce a much more efficient microcavity. However, silver has a significantly higher material cost than aluminum.

Traditionally, true Fabry-Perot cavities were cavities formed between two planar reflectors which permitted a small portion of the energy to escape. In this configuration, at angles other than the angle normal to the plane of the electrodes, the resonant wavelength of the light changes as a function of angle, resulting in a change in the emitted color of light as a function of viewing angle. Further, it is desirable to tune the resonant frequency of the cavity to match the peak wavelength of the emitter to gain maximum efficiency. For angles away from normal, the resonant frequency of the microcavity will not match the peak wavelength of the emitter. As a result, the luminance output by the device will be less than the luminance output in the direction normal to the cavity. Therefore, changes in the luminance output of the device will accompany changes in color as a function of viewing angle. However,

6.1 Organic Light-Emitting Diode (OLED) Displays

in many modern devices the reflectors are not uniformly parallel to the other. In well-designed devices of this type, the color can be maintained as a function of viewing angle and the luminance loss with viewing angle can be equalized between the channels. In this configuration the relative energy between the color channels is maintained across viewing angles and little, if any, change in chromaticity coordinates or perceived color are present. Instead, the device emits less energy as the viewing angle increases resulting in luminance loss with increased viewing angle. However, without a reference, the human visual system is not sensitive to modest changes in luminance and therefore this luminance loss is not apparent.

In summary, while OLED materials were originally challenged in providing highly saturated colors, improvements in materials and device structure have largely overcome these deficiencies. Modern OLED devices are regularly able to provide highly saturated primaries, often meeting or even exceeding the P3 color standard. Further, well-engineered displays provide little, if any, chromaticity variation with viewing angle. Some OLED displays will exhibit luminance variation as a function of viewing angle. However, this variation is not typically visible to the human eye due to luminance adaptation.

6.1.4 Spatial Distribution and Reflectance

Numerous attributes of an OLED make light extraction difficult. The fact that OLEDs produce Lambertian emission and that one electrode is typically semi-transparent and one electrode is typically opaque, implies that less than half the light will be emitted in a direction toward the transparent electrode where it could be viewed by an observer. Further, light emission occurs in a medium with a relatively high index of refraction as the index of refraction for OLED materials is often in the neighborhood of 1.7. As a result, using a simple Snell's Law calculation and assuming that the light must pass from a material with an index of refraction of 1.7 into air, no more than the light emitted within a 72° cone will escape from the device, which simple geometric calculations will indicate, permits about 15% of the light to escape from the device. This further assumes that all of the materials in the device are fully transparent. Although this relatively simple calculation ignores the fact that there are actually multiple changes in index of refraction in a typical OLED device due to small differences in the index of refraction between OLED materials themselves as well as the ITO and large changes in the index of refraction between these materials and glass, this relatively simplistic calculation indicates the magnitude of the issue resulting from Lambertian emission within a high index medium.

A common method to improve efficiency is to utilize a reflective electrode. This electrode reflects the light traveling towards the back of the display, directing it towards the front of the display. Depending upon the reflectance of the electrode and the opacity of the OLED materials between the light-emitting layer and the electrode, one could reasonably expect somewhere between a 60 to 100 percent improvement in light extraction efficiency if all of the surfaces were perfectly flat. However, the surfaces are not perfectly flat and therefore this reflective electrode can result in efficiency improvements greater than 100% as light that would otherwise be trapped within the device would appear to be redirected and permitted to escape after multiple reflections.

> In an OLED display, the electrode behind the OLED emitter is usually formed from a highly reflective metal, which can significantly improve display efficiency. Unfortunately, this electrode also reflects ambient light. This reflectance is typically overcome by applying a circular polarizer to the front of the display.

Although the use of a reflective electrode improves the efficiency of light emission from the OLED, it has a negative implication for ambient light reflectance of an OLED display. Light which enters the display from the ambient environment will be reflected from the electrode and returned to the user. Further, the backplanes that are typically constructed for OLED displays also include electrical components (i.e., metal power and drive lines, TFTs and capacitors) that are highly reflective. As a result, any OLED display without further optical components will have a specular reflectance that can exceed 50%. Numerous approaches can be used to overcome this problem.

We have already discussed the use of microcavities to improve the purity of the color of light emission from an OLED display. However, microcavities have the additional benefit that they trap ambient light that does not have a wavelength near an integer multiple of the microcavity thickness. Therefore, microcavities can reduce the reflectance of the OLED, particularly to wavelengths of light that are not near the emission wavelength of the light emitting element.

A common approach to reduce the reflectance of an OLED display is to apply a circular polarizer to the front of the display. Light which passes from air through the circular polarizer is given a right or left handed circular polarization. This polarization direction is reversed on reflection from the reflective electrode. When this reflected light encounters the circular polarizer at the front of the display, the polarization of the light is now opposite to that which the polarizer transmits and the light is absorbed. Circular polarizers are produced with various levels of quality. The highest quality circular polarizers with a front surface diffuser can reduce the reflectance of the display (specular or diffuse) to only about 0.2%, but also absorbs 60% of the light emitted by the display. Lower quality films are less expensive but will result in substantially higher reflectance (on the order of 4%) and absorb 40–50% of the light emitted by the display.

Color filters are often applied in certain OLED architectures. These filters absorb most of the light outside their color band. For example, a blue color filter will absorb most green and red light, as well as a little bit of blue light. Further, the ambient light within the passband of the filter must pass through the color filter twice, once when it enters the display and once when it exits. Therefore, color filters can reduce the reflectance of the light significantly even within this passband.

6.1 Organic Light-Emitting Diode (OLED) Displays

Each of these techniques for reducing the reflectance of the display are commonly used and are complimentary. That is they each reduce the reflectance of the display through a different means and light which might be reflected if only one of these techniques was applied is further reduced if a second technique is applied in the same display. The combination of a circular polarizer together with one or more of the other techniques results in a display with very low reflectance, such that there is very little light reflected. This coupled with the ability to turn off light emitting elements in the display, so that the display emits no light from these regions, results in very high contrast displays.

6.1.5 Lifetime

An attribute of any practical light emitting device is the loss of efficiency which occurs as a function of use. Both iLEDs and OLEDs suffer from this same phenomena. In either of these types of devices, the output of the device for a constant voltage decreases as a logarithmic function of the current density within the device. Current density refers to the ratio of the current to the area of the device. In most devices, this is not a significant issue as the entire device loses luminance at once simply becoming dimmer with time. Since our eyes are not sensitive to absolute luminance, this loss is inconsequential until very large luminance losses occur. For example, the florescent lamps within all LCDs become less efficient and dimmer with time but because the bulbs all age evenly and our visual system is relatively insensitive to changes in absolute luminance, these changes are imperceptible. However, in LED technologies, each device (i.e., each light emitting element in an OLED) can have different amounts of energy applied to them and therefore, each device will age at a different rate. This localized aging results in an artifact referred to as "burn in". Additionally, different colors of light-emitting materials can age at different rates. This difference in lifetime can result in unwanted color shifts with time.

Burn in is a particular problem in OLED displays as individual OLEDs form individual light-emitting elements. Therefore, activating one light-emitting element with a high current for an extended period of time without activating its neighbor causes the active light-emitting element to lose efficiency faster than its neighbor. If these light-emitting elements are later turned on with the same voltage or current signal, the previously active light-emitting element will be less efficient and therefore have a lower luminance than its previously inactive neighbor. If these neighboring OLEDs output the same color of light, the result is an unintended high frequency pattern to which the human eye is sensitive. If these two neighboring OLEDs output different colors of light within a full color display, there will be less than the intended color of light output by the previously active light-emitting element. Thus, an unintended color variation will exist. Again, the human eye can be quite sensitive to this unintended local variation in color.

> Lifetime of OLED materials are sufficient to provide displays which output light over decades of display use. Unfortunately, individual OLEDs within a display may age at different rates. As this non-uniform aging occurs, some light emitting elements become dimmer than their neighbors. This can result in the perceived artifact referred to as "burn in" which appears as low luminance areas or color shifts within small areas of a display. Without proper amelioration techniques, this artifact can occur when icons are displayed in fixed locations on the display for long periods of time.

In OLED displays, this problem is exacerbated by the fact that different colors of light emitting materials can lose efficiency faster than others when exposed to the same current density. For example, blue light-emitting materials often lose efficiency faster than green light-emitting materials which lose efficiency faster than red light-emitting materials when exposed to the same current density for the same period of time.

Although less of a problem, it is also worth noting that this phenomena does not affect only the OLED device within a common OLED display. In these displays, thin-film transistors control the flow of current to each OLED. Similarly to the OLED, these transistors can experience stress under high current, which reduces their efficiency, reducing the amount of current which passes through to the OLED. For most backplane technology, the TFTs are relatively robust and this loss of efficiency occurs much more slowly for the TFT than the OLED. However, this phenomena can be a significant problem, particularly in low-cost TFT backplanes, for example those formed from amorphous silicon (aSi).

As a result of lifetime issues, some light emitting elements can appear dimmer or different in color than neighboring light emitting elements, resulting in localized visual artifacts. These artifacts are often referred to as "burn-in" artifacts. Additionally, the color of OLED displays can shift towards red or yellow as the blue light-emitting elements lose efficiency faster than the red or green light-emitting elements. Each of these artifacts can be influenced through display design, including through the use of purposefully chosen multi-primary display structures. Additionally, they can be influenced by image processing techniques that perform functions like dimming the display when static images containing high current density regions are displayed or providing small spatial shifts in the image content on the display over time to reduce the spatial frequency of the boundaries of the burnt in regions which usually reduces the visibility of these artifacts.

It should be noted that these artifacts will be common between OLED and iLED displays. However, the lifetime of iLED materials is often longer than the lifetime of OLED materials. Therefore, depending upon the current densities which are applied in the final device, these lifetime artifacts may be less problematic in iLED displays than in OLED displays.

6.1.6 Display Structures

Two general OLED display structures are currently being applied in production today. These include a color patterning approach and a white with color filter approach. Each approach has significant advantages and disadvantages. However, the largest distinguishing characteristics today are power consumption and ease of manufacturing. The color patterning approach produces displays with lower power consumption, thus making this approach desirable in displays for portable electronics. The white OLED with color filter approach is not as power efficient as the color patterned display, but is much easier to produce, especially for large area displays. Thus, this approach is primarily applied for the television and advertising display markets.

6.2 Color Patterning

The color patterning approach applies a display structure similar to the one shown in Fig. 6.4. As shown in this figure, the light-emitting layer within the OLED is patterned, permitting each light-emitting element to produce either red, green or blue light. These light emitters are typically placed within a microcavity to improve the color saturation of the emitted light. Thus, the materials in the light-emitting layer and the size of the corresponding microcavity dictate the spectrum of the resulting light.

Importantly, this approach requires an additional manufacturing step to produce the microcavity and an additional manufacturing step to lay down each color of emitter. In traditional OLED manufacturing, the OLED materials are deposited on the display substrate through a process known as thermal evaporation. The temperature of the substrate is maintained at a lower temperature and suspended over a container of OLED materials. The OLED materials are then heated until they evaporate. The plume of evaporated materials then rises and condenses onto the substrate to form a layer of OLED materials on the substrate. This process is very reliable when depositing materials over the entire display substrate as the rate of evaporation can be controlled by controlling the temperature of the OLED materials. However, the direction of the plume of evaporated material cannot be controlled and therefore the plume tends to cover large areas of the substrate.

To form a color display, deposition of the light-emitting layer must be highly controlled such that blue light emitting materials are deposited only over the blue emitting light-emitting elements. A way of achieving this is through the use of a shadow-mask. A shadow mask is normally formed from material such as Inconel. This shadow mask has one small hole in it for each blue light-emitting element. During manufacturing, this shadow mask is precisely aligned with the substrate allowing the holes in the shadow mask to be aligned with the blue light-emitting elements on the display. The blue light-emitting materials are then deposited by vaporizing material below the shadow mask. As the vaporized material encounters the shadow mask most of the light-emitting material lands on the shadow mask and is deposited there. However, a portion of the blue light-emitting material passes through the holes in the shadow mask and is deposited on the blue light-emitting elements of the display substrate. The shadow mask is then removed. A similar green shadow mask is then aligned to the display substrate and green light-emitting materials are deposited. The green shadow mask is removed and the process is completed again to deposit the red light-emitting materials onto the red light-emitting elements.

> Forming patterns of differently-colored light emitting elements presents a significant challenge in OLED manufacturing. This challenge becomes especially significant when forming large displays.

While generally effective, this approach suffers from a number of inefficiencies. First, the precision alignment of the shadow mask is difficult. Second, the shadow mask can be heated during deposition of the materials and thermal expansion of the shadow mask can alter the alignment, often resulting in misalignment of the shadow mask for some areas of the substrate. Third, most of the light-emitting materials are actually deposited on the shadow masks. This not only wastes a large proportion of one of the most expensive materials in OLED production, but also requires additional steps to clean the shadow mask before it is reused. It is also important that the thermal expansion of the shadow mask increases as the area of the shadow mask increases. This approach becomes more difficult and expensive as the size of the display to be produced increases, making this the preferred approach when manufacturing small, portable displays. In fact, most OLED displays being produced today for use in cellular telephones are produced through a similar method.

Many replacement methods to this technique have been explored. However, the leading candidate today is to print these light-emitting layers instead of evaporating them. Although significant investments have been made in this area, printing techniques are only beginning to reach mass production. For example, the OLED display released in the fall of 2017 in the Google Pixel 2 XL has been reported to utilize a printed OLED display. Unfortunately, it has been reported in the media that this display exhibits "burn in" issues [1]. Although not publicly discussed, it is possible that the current printing techniques are still not capable of producing high enough quality OLEDs to be applied in the mass market.

6.3 White OLED

An alternative to the color patterning approach is the formation of an OLED through formation of a white-emitting OLED and color filters. An example architecture can be formed as shown in Fig. 6.10. As shown in this figure, the bottom substrate can be formed to permit the formation of microcavities. An initial OLED is formed on this

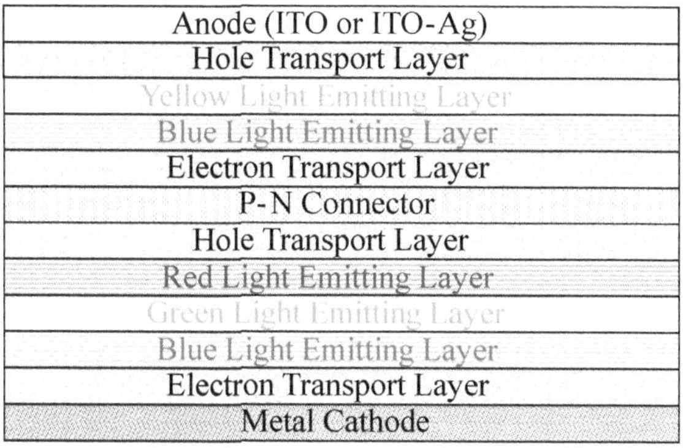

Fig. 6.10 Illustration of a tandem white emitting OLED useful in constructing an OLED display using white emission with color filters

Anode (ITO or ITO-Ag)
Hole Transport Layer
Yellow Light Emitting Layer
Blue Light Emitting Layer
Electron Transport Layer
P-N Connector
Hole Transport Layer
Red Light Emitting Layer
Green Light Emitting Layer
Blue Light Emitting Layer
Electron Transport Layer
Metal Cathode

substrate. This initial OLED can include multiple light-emitting layers, for example it might include a blue, a red, and a green light-emitting layer. Once the initial OLED is formed, a connector layer might be coated on the first OLED. A second OLED can then be formed on top of the first OLED, where this second OLED also has multiple light-emitting layers, for example a blue and a yellow light-emitting layer. By forming an OLED structure with multiple light-emitting layers, the OLED emits light having multiple emission peaks (e.g., red, green, blue, and yellow), producing an overall white light [15]. The amplitude of each of these peaks is controlled through changes in the deposition thickness of each of the light-emitting layers. This OLED can then be paired with a cover glass that is coated with red, green, and blue color filters to form a full color display.

In this structure, it is not necessary to pattern any of the OLED materials. Therefore, the use of the fine metal shadow mask, which presents manufacturing difficulties for the color patterning approach, is avoided. This structure does require more OLED materials to be blanket-coated across the display. Therefore, it requires an increased number of simpler OLED material deposition steps. Additionally, this method requires the formation of a color filter array which is similar to the color filter array formed in the manufacture of LCDs. The necessity of the color filter requires additional, well understood manufacturing steps and additional material costs. Further, a significant amount of the light produced by the OLED materials is absorbed by the color filters, making this structure less energy efficient.

The use of color filters can eliminate the need to pattern different colors of OLED materials during manufacturing, simplifying the manufacture of these displays. However, this approach can significantly reduce the power efficiency of the resulting display due to the absorption of the color filters.

As this structure does not require the use of fine metal masks, which limits the ability to form large OLED displays using color patterning, this approach is more amenable to the production of large displays. As a result, this approach has been applied predominantly by LG Display in the production of consumer television, an application where the constant availability of power makes the power efficiency of this device structure less troublesome. It should, however, be noted that to improve the power efficiency of these displays, LG display does not employ only red, green, and blue light-emitting elements in their display, but instead the display and color filters are designed to produce an additional white or unfiltered light-emitting element within the display, making this display one of the only mass-produced, direct-view, multi-primary displays available in the consumer marketplace.

An important element of the OLED structure is the use of multiple OLEDs which act in serial to form a tandem OLED. In this structure, each OLED formed in serial on the substrate produces an amount of light as a function of current through the OLED. Therefore, for a given current, this tandem structure produces approximately twice the light output of a traditional OLED. As these devices are in series, the voltage required increases, apparently eliminating any gains in power efficiency. However, this tandem structure has a number of additional potential advantages. As mentioned earlier, the lifetime of the OLED is dependent upon the current density through the OLED. To produce a desired luminance level, this structure effectively halves the current, reducing the current density through the OLED and extending its lifetime. Therefore, the lifetime of tandem OLEDs are typically much greater than single-layer OLEDs. As we will see shortly, the reduction in current within an OLED also decreases the power consumption of an OLED display through decreasing IR drop within the display, another important element.

Although alternate display structures have been explored, the patterned RGB and un-patterned, color filtered RGBW OLED displays are the primary structures in mass production. Combinations of these display structures have also been attempted. For example, SONY Corporation currently markets an RGB patterned OLED display, which includes color filters for the cinema market. However, this display is not being produced in large numbers.

6.3.1 Other Power Considerations

We have discussed the structure of OLEDs without much discussion of the active matrix display itself. Similar to LCD, modern OLED displays are formed on active matrix substrates where data is provided to each light emitting element through a thin film transistor to control the light provided by each light emitting element. The requirements for backplanes in OLED displays are, however, significantly different than the backplanes for LCDs. As we discussed earlier, the backplane in an LCD must deliver voltage to the electrode of an LCD to establish an electric field within the liquid crystal cell. As such, current does not flow through the liquid crystal, but the liquid crystal reacts to the presence of the electric field. However, as we have discussed, light output of an OLED is modulated by modifying the current which flows through an OLED. Therefore, in an OLED display, the backplane must provide current to the OLED and the complimentary electrode must carry current away from the OLED. Further, the Thin Film Transistors (TFTs) must be capable of reliably regulating current uniformly across the substrate without degrading significantly with time.

It is important that the need to provide current places additional requirements on the backplane and the materials from which TFTs are made. LCDs often employed amorphous silicon (a-Si) as a semiconductor within the TFTs, unfortunately these transistors have low mobility, restricting the rate of current flow through the TFT. Further, transistors made from a-Si degrade relatively rapidly with time when exposed to high currents, reducing the amount of current to the OLED, which enhances the appearance of burn in. High end LCDs often employed Low Temperature Polysilicon (LTPS). LTPS is formed by depositing amorphous silicon and then heating the silicon with a laser to cause it to crystalize. Unfortunately, this process does not result in a uniform layer of semiconductor and therefore the mobility of the semiconductor is not uniform, typically resulting in a substrate where some rows of the TFTs within the display have a higher mobility than other rows. As a result, more current flows through some rows of TFTs than others and the brightness of the OLEDs vary across rows of the display, often resulting in a very apparent striped pattern. Image processing has been applied to improve the quality of OLED displays formed with either of these two types of backplanes [3, 7]. To avoid this complication, the industry has sought improved semiconductor materials for use in OLED display manufacturing, including the use of new materials such as doped zinc oxides [10] and new ways of crystalizing silicon to form materials that are more like single crystal silicon than LTPS [4]. However, the mass production of stable, highly uniform backplanes provides a continuing challenge to the industry.

The fact that OLED displays must provide current to each light emitting element creates additional constraints for the OLED substrate. Figure 6.11 shows an illustration of a backplane. Similar to the portion on the LCD backplane shown in Fig. 5.2, this substrate includes an electrode, select lines and data lines. It also contains a TFT which permits a signal from a data line to load a voltage onto a capacitor, labeled select TFT. However, an OLED typically has at least one additional TFT, denoted as the power TFT. The power TFT is connected to the capacitor. The voltage on the capacitor then switches on the power TFT according to the voltage stored in the capacitor and in response to this voltage, the power TFT regulates the flow of current from a power line to the electrode. This power line then provides current to each OLED through each electrode. As a result, the power line must be capable of carrying current to each OLED along the height of an OLED display. This power line is also manufactured in a thin film deposition process and is therefore quite thin. As a result, the power line has a significant resistance. There is, therefore, a loss of voltage along the power line, calculated according to the following equation:

$$V = IR \qquad (6.3)$$

where V is the voltage, I is the current in amps and R is the resistance in Ohms. This drop in voltage then results in a power loss along the power line, according to Eq. 6.4, shown as:

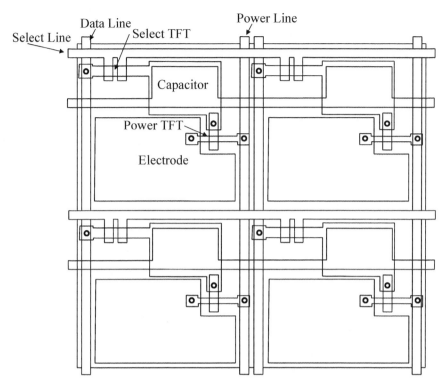

Fig. 6.11 Illustration of a portion of an example OLED substrate. Note that the circles indicate vias, permitting connection between patterned top and a bottom metal layers where these metal layers are separated by a semiconductor, such as LTPS. The electrodes may be formed from a metal oxide, such as ITO in bottom-emitting OLED displays and a metal such as aluminum in a top-emitting OLED display

$$P = I^2 R \qquad (6.4)$$

where P is the power loss in Watts and I differs along the length of the power line as current is delivered to each light-emitting element.

As a result, the power loss across the power line is a function of the square of the current to each OLED. The electrode on the opposing side of the OLED also has a resistance and a similar power loss can occur through this electrode. While this power loss is small between any two light emitting elements in a display, it becomes significant for large displays, which have long power lines that must provide power to large numbers of OLEDs. Note that it is also possible for drops in voltage to occur down the length of the power line which reduces the amount of current flowing to OLEDs towards the end of the power line, which can produce luminance non-uniformities and artifacts. It is important that this effect is highly dependent upon the current which is required by the OLED. A significant advantage of the tandem OLEDs used by LG in their televisions is that twice as much light is produced for a given current or, stated another way, for a given amount of light, the current is halved. Therefore, the use of tandem structures can significantly reduce the power loss through the power lines of the display.

It should also be noted, that while the substrate shown in Fig. 6.11 contains only 2 TFTs, this is the minimum number of TFTs required to drive an OLED within an OLED display. Some designs require additional TFTs to attempt to compensate for OLED burn-in, loss of TFT mobility, or other attributes of the display.

When discussing an LCD, we discussed the fact that increasing the area devoted to items other than the electrode reduced the efficiency of the LCD as light could not travel through these elements. These elements have a different effect in an OLED. The efficiency of the OLED is dependent upon the OLED and the optical structure of the OLED, with light output controlled by the amount of current passing through the OLED. The area of the electrode governs the area of the OLED through which current flows. That is, it affects the current density, or current per unit area, of the OLED. The current density is related to the lifetime of the OLED with linear increases in current density causing an exponential loss of OLED lifetime. So reducing the relative area of the electrode reduces the lifetime of the OLED but has no effect on efficiency. Further, in top emitting OLED designs, where the electrode on the substrate is formed from a reflective metal, it is possible to design and build substrates where the supporting electrical structure (select, data and power lines, as well as TFTs and capacitor) is constructed on a different plane of the backplane than the electrode. While such a design would add manufacturing steps and cost, it is theoretically possible because the light which is produced does not pass through the substrate.

> The need to provide known amounts of current to each OLED in a display significantly complicates the design of the OLED backplane as compared to LCD displays. Further, the loss of power due to the resistance of power lines in a backplane can decrease the power efficiency of OLED displays, especially for large displays.

6.3.2 OLED Summary

In this section, we have discussed the construction of displays using OLED technology. Unlike LCD, where the display technology modulates the passage of light from a backlight to an observer, OLED displays create light only where it is desired. As a result, the display technology can create a display with nearly infinite contrast and consumes little power when displaying dim images. Highly energy efficient OLED displays can be formed by patterning emitting materials, permitting the creation of displays with arrays of highly saturated red, green, and blue light emitting elements. This structure provides especially high efficiency and high color saturation when employing the optical effects of micro-cavities. However, existing manufacturing techniques make it difficult to reliably pattern OLED materials over large substrates. As a result, patterned RGB OLED displays are typically limited to relatively small displays. OLED displays may alternately be manufactured by producing un-patterned OLEDs which emit light with a broad bandwidth and employ color filters to create color differentiation. While this technique can produce displays of practically any size, these displays will typically trade power consumption and color saturation through color filter design and will not be as power efficient as color patterned OLED displays. Finally, we discussed the fact the OLED displays require significantly higher integration than LCDs with the manufacturer of the backplane providing a much larger portion of the value added know-how and manufacturing investment necessary for production of the display.

Although OLED displays have many advantages, they are complex to construct and suffer from image artifacts such as burn in, which is difficult to eliminate. Limitations in lifetime due to burn in and manufacturability provide the primary barriers to broader adoption of the technology.

6.4 Inorganic Light Emitting Diode Displays

As we saw in the preceding section, OLED displays have a number of desirable attributes, including creation of saturated light only where needed within a display. As a result, the displays hold the promise of creating high quality, highly energy efficient displays. Unfortunately, manufacturability and lifetime concerns limit the adoption of this display technology. As we discussed earlier iLEDs have many desirable qualities for light creation. Therefore, one might ask: "Is it possible to form displays from inorganic LEDs?" If such a display could be manufactured, it would provide the desirable attributes of an OLED display without the lifetime limitation.

Unfortunately, current methods for manufacturing iLEDs involve beginning with a crystalline wafer made from materials such as gallium arsenide or gallium nitride and then growing crystalline iLED wafers on top of these substrates. These wafers are limited in size, typically to wafers that are 5–8 in. in size. Further, the wafers and the process is relatively expensive such that making a single display on a wafer would be cost prohibitive. Displays can, however, be formed by cutting the waver into small pieces and distributing these pieces across a substrate. These individual iLEDs can then be connected to drive electronics to form a display. In fact, electronic billboards have been formed through robotic or manual labor by placing millions of iLEDs on a large substrate to form very large displays. The question then becomes, could such a process be scaled cost effectively to smaller displays, displays that are appropriate for consumer television or smaller.

Recently techniques have been developed which permit a sacrificial layer to be grown on the crystalline wafer and the iLED to be grown on top of the sacrificial layer. After the iLED is grown, the iLED is cut into very small chiplets and the sacrificial layer is partially removed [5]. Finally a polymer stamp is used to pick up sparse arrays of the iLEDs and transfer them onto a substrate with space between them. Through this pickup and transfer process, red, green, and blue iLEDs can be deposited onto a larger substrate [11]. Similar processes can be used to deposit small chiplets of crystalline silicon which contain drive circuits for the iLEDs. However, techniques have been demonstrated for producing these circuits on the same chiplet as the iLED [17]. Alternate technologies have been discussed which use roll to plate deposition techniques for both the circuitry and iLEDs. This technology is claimed to be extensible to any sized display, permitting the production of microdisplays through larger billboards [6].

Through one of these processes, high resolution displays can theoretically be constructed by printing iLEDs. Such a display would have many of the desirable properties of OLEDs without the lifetime concerns. The primary drawback of this technology is the formation of high quality displays at acceptable cost. Unfortunately, as this approach requires each light emitting element to be transfer printed, the likelihood of defects within the display is extremely high. As a result methods are being sought to reduce the likelihood or perceived quality loss due to defective light emitting elements. If this issue can be overcome, this technology could provide an alternative high quality display technology.

> The ability to form electronic displays from iLEDs has the potential to overcome many of the challenges presented by OLED displays. Unfortunately, the requirement to transfer print millions of iLEDs without any visually apparent defects is difficult to achieve and will likely limit the adoption of this technology.

6.5 Summary and Questions for Thought

In this chapter, I have discussed OLED display technology, which is the current market contender, and transfer printed iLED display technology, which is a potential future display technology. OLED or iLED technologies have the potential to create light only where needed and therefore excel in these same areas. However, each of these technologies have significant manufacturing hurdles to overcome. Despite the challenges very high quality, highly integrated OLED displays are being produced in ever increasing numbers in both the small handheld and larger television markets. The current production rates for OLEDs are significantly less than the production rates for LCDs. OLED displays, are, however, increasingly considered a premium display technology, providing potential consumer differentiation over LCDs.

Although not discussed earlier, OLED display technology has been discussed as having other desirable characteristics, such as the ability to be printed on transparent substrates permitting transparent displays, the ability to be printed on flexible substrates, creating flexible displays, and the ability to utilize smaller and smaller light emitting elements without loss of power efficiency, potentially enabling multi-view displays. This has forced the LCD community to explore responses to these potential challenges. These potential innovations as well as the natural higher quality inherent in OLED display technology has created innovation across the entire direct view display space that would not exist without the competitive challenge presented by OLED technology.

Looking back at this chapter, a few questions for thought can be proposed:

(1) What is the comparative value of highly integrated verses loosely integrated technologies? When is it useful to have a large number of companies contributing to the development of underlying loosely integrated technologies verses fewer companies contributing to a highly integrated technology?
(2) Given the differences in contrast and viewing angle between LCD and OLED, how important are these two attributes within the consumer market place?
(3) How does ambient viewing conditions affect the value of contrast in a display system?
(4) The industry has targeted OLEDs predominantly at the handheld and television markets. If burn-in is a significant issue for OLED technology, are these the best markets for entry? Why or why not?
(5) If printing OLED display emitters is not practical, can white OLED with color filters challenge LCD in markets beyond television?

References

1. Baig EC (2017). Pixel 2 XL smartphone "burn in" could burn Google. https://www.usatoday.com/story/tech/columnist/baig/2017/10/26/pixel-2-xl-smartphone-burn-in-could-burn-google/796982001/
2. Burroughes JH, Bradley DDC, Brown AR, Marks RN, Mackay K, Friend RH, Holmes AB (1990) Light-emitting diodes based on conjugated polymers. Nature 347(6293):539–541 https://doi.org/10.1038/347539a0
3. Chaji GR, Alexander S, Nathan A, Church C (2008) Low-cost amoled television with IGNIS compensating technology. In: Proceedings of SID, pp 1219–1222. https://doi.org/10.1889/1.3069355
4. Chaudhari K, Chaudhari A, Chaudhari P (2014) Method of growing heteroepitaxial single crystal or large grained semiconductor films on glass substrates and devices thereon. United States Patent Number 8,916,455
5. Greenemeier L (2009) Brighter idea : next-generation inorganic LEDs promise longer lives and more lumens. Sci Am
6. Hildebrand N (2017) How Korea wants to become a leader in micro LED technology. Display Daily. https://www.displaydaily.com/index.php?option=com_content&view=article&id=57251:how-korea-wants-to-become-a-leader-in-micro-led-technology&catid=152:display-daily
7. Kohno M (2010) Assuring uniformity in the output of an OLED. United States Patent Number 7,859,492. https://doi.org/10.1197/jamia.M1139.Adar
8. Kondakov DY, Pawlik TD, Hatwar TK, Spindler JP (2009) Triplet annihilation exceeding spin statistical limit in highly efficient fluorescent organic light-emitting diodes. Journal of Applied Physics 106:124510

References

9. Kwong R, Ma B, Xia C, Alleyne B, Books J (2013) Phosphorescent materials containing iridium complexes. United States Patent Number 8,431,243
10. Lee H-N, Kyung J, Sung M-C, Kim DY, Kang SK, Kim S-J, Kim S-T (2008) Oxide TFT with multilayer gate insulator for backplane of AMOLED device. Information Display 16(2):265–272. https://doi.org/10.1889/1.2841860
11. Meitl MA, Radauscher E, Rotzoll R, Raymond B, et al (2017) Emissive Displays with Transferred Microscale Inorganic LEDs. In: SID symposium digest of technical papers, pp 257–263, Los Angeles, CA
12. Murano S, Kucur E, He G, Blochwitz-nimoth J, Ag N, Hatwar TK, Van Slyke S (2009) White fluorescent PIN OLED with high efficiency and lifetime for display applications. In: Proceedings of SID, pp 417–419. Los Angeles, CA
13. Seo S, Sasaki T, Sugisawa N, Nowatari H, Ushikubo T, Oe Y, Seo PS (2010) High-efficient green OLED over 150 lm/W with New P-doped layer exhibiting no optical loss derived from charge transfer complex. In: SID symposium digest of technical papers, pp 1804–1807
14. Spindler JP, Begley WJ, Hatwar TK, Kondakov DY (2009) High-efficiency fluorescent red- and yellow-emitting OLED devices. In: Proceedings of SID, pp 420–423. Los Angeles, CA
15. Spindler JP, Hatwar TK, Miller ME, Arnold AD, Murdoch MJ, Kane PJ, Van Slyke SA (2006) System considerations for RGBW OLED displays. Journal of the Society for Information Display 14(1):37. https://doi.org/10.1889/1.2166833
16. Tang CW, VanSlyke SA (1987) Organic electroluminescent diodes. Appl Phys Lett 51:913. https://doi.org/10.1063/1.98799
17. Tull BR, Twu N, Hsu Y, Leblebici S, Kymissis I, Lee VW (2017) Micro-LED microdisplays by integration of III-V LEDs with silicon thin film transistors. In: SID symposium digest of technical papers, pp 246–248. Los Angeles, CA

Chapter 7
Display Signal Processing

7.1 RGB Display Rendering

We will begin a discussion of display rendering by discussing a relatively standard image processing path for a RGB display. This discussion will include the steps shown in Fig. 7.1. As shown in this figure, we must first select a display white point. We will then decode the image data into a space where the code values are linearly related to scene luminance. If the white point of the encoded image is different than the selected white point, we will need to transform the image data so that the colors within the image with the white point selected in step 1 appears similar to the colors in the initial scene. To progress further, it is necessary to describe the performance of our target display. This data is then used to derive a color matrix which permits us to render the image to the space provided by the display primaries. Finally, we will need to compensate the image for any nonlinear relationship between the display code values and the luminance of the display.

It is noteworthy that steps 1 and 4 are performed for any display. However, the other steps need to be completed for each pixel of image data input to the display. It is also noteworthy that these steps must be performed in a way that avoids significant loss of data. While the image data may be encoded in a 8-bit image format and the final display will likely accept only 8 bits per code value, the steps performed between 2 and 6 inclusive will likely need at least 10 bit processing and may require more as this processing is performed in a space that is linear with luminance. This is relevant since, as we discussed earlier, the human visual system is much more sensitive to small errors in luminance for low luminance values than for high luminance values.

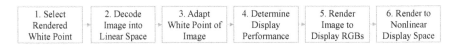

Fig. 7.1 Overview of display rendering process

Importantly, this is a basic image processing path that seeks to render images as a metameric color match to the original scene. Additional steps may be desirable to support further image enhancements. Finally, this image processing path is described as it will support addition and subtraction of values without color distortion, which will be important as we begin to discuss transformation of RGB images to more than 3 channel images, which are useful for driving multi-primary displays.

> Performing color manipulation in a space where code values are linear with scene and display luminance permits one to maintain a metameric match between the displayed image and the original scene. However, image processing in this linear space will require 10 or more bits to maintain appropriate image quality.

7.1.1 Selection of Rendered White Point

As we discussed earlier, most images are encoded with the assumption that they will be presented on a display having a white point equivalent to D65. That is, the display is assumed to have a color temperature of 6500 K. In many electronic displays, including LCDs with LED backlights, it is possible to drive the display to a bluer white point, such as 9300 K. These bluer white points are often preferred by consumer display manufacturers as the display with a bluer white point will appear brighter to a user for a given power consumption than the display would appear if the display had a white point of D65. Therefore, it is standard practice within the display industry to balance the primaries of the display to have a default white point that is bluer than D65. In fact, many consumer displays have default white points that are bluer than 9300 K and often have white points of 11,000 K or greater. At the same time, television standards assume D65 and therefore most televisions come equipped with a "Cinema" mode, which adjusts the primaries of the display to provide a white point nearer D65. In the current discussion and example, we will assume that our desired white point is 9300 K. This permits us to illustrate the use of a chromatic transform within our image processing path.

7.1.2 Linearization and Decoding

The second step in the process is to decode the image into a space where a change in code value is linear with a change in rendered scene luminance. In this step we seek to linearize the data with respect to scene luminance. This step is important as it permits one to scale the color channels while maintaining the relative relationship

7.1 RGB Display Rendering

between the luminance values in each channel, as well as to subtract luminance from one color channel and add it to another. Once the values are linearized, we can scale the color channels using multiplication or division and the relative relationship between luminance values achieved between any code value and the maximum code value will be maintained. In a more specific example, 50% of the luminance will be 50% of the luminance before or after the value is multiplied or divided. Assuming that the input image data is encoded as sRGB component values, the original scene luminance values have undergone the nonlinear encoding step depicted in Eq. 4.1. This encoding step will then be reversed using the following equation:

$$C_{linear} = \begin{cases} \frac{C_{sRGB}}{12.92}, & C_{sRGB} \leq 0.04045. \\ \left(\frac{C_{sRGB}+a}{1+a}\right)^{2.4}, & C_{sRGB} > 0.04045. \end{cases} \quad (7.1)$$

where C_{linear} represents the linearized code value and "a" equals 0.055, as defined in the sRGB standard. This computation is completed independently for all RGB code values. The resulting linearized code values can then be converted into CIE XYZ tristimulus values through the following matrix operation:

$$\begin{bmatrix} X \\ Y \\ Z \end{bmatrix} = \begin{bmatrix} 0.4124 & 0.3576 & 0.1805 \\ 0.2126 & 0.7152 & 0.0722 \\ 0.0193 & 0.1192 & 0.9505 \end{bmatrix} \begin{bmatrix} R_{linear} \\ G_{linear} \\ B_{linear} \end{bmatrix} \quad (7.2)$$

Through this decoding step, we obtain an approximation of the original scene XYZ values, which can be accurately rendered to the display for viewing. It should be noted, however, that the original encoding step did lose some information and colors which were outside the boundary of the sRGB color space have been clipped such that they now lie on the boundary of the encoded color space. Of course images encoded in other color spaces would undergo an analogous decoding to some known, common color space, such as XYZ tristimulus values. Because the capture system attempted to expose the sensor properly, much of the scene luminance information has been removed from the image data and the resulting XYZ values are normalized to a range between 0 and 1.

It may also be important to recall that most image encoding standards define the location of a diffuse white to be less than the maximum white value. For sRGB, a diffuse white is assumed to have code value 235. Values above 235 are assumed to contain highlight information, such as specular highlights that might occur due to reflection of the sun in a mirror or other smooth surface.

Table 7.1 Tristimulus values for D65 and 9300 K white points

White point (K)	X	Y	Z
6500	95.043	100	108.890
9300	95.286	100	141.41

7.1.3 Adapting the Image Data to an Alternate White Point

At this stage, we have linear XYZ values for the scene with an assumed D65 white point. However, we wish to render these images on a display with a target white point of 9300 K within our current example. Note that the human eye will adapt to each of these white points. Therefore, we have to account for the difference in cone response for the adapted white point. That is we must account for the differences in human cone response between the two adaptation points. This process generally consists of transforming the tristimulus values for the image into cone response values, then adjusting these cone response values to adjust for a change in eye adaptation, followed by converting the images back to tristimulus values. Perhaps the simplest of these procedures performs the adaptation process through linear scaling of the cone responses. One such method performing this adaption is the use of von Kries transformations [3].

> When the white point of an input image is different than the white point of the final displayed window, it is necessary to apply a color adaptation transform to account for differences in the adaptation of the human visual system.

We begin with the tristimulus values shown in Table 7.1 for the two white points, assuming that the adapting luminance (Y) is set to 100. The cone responses are calculated as shown in Eq. 7.3. This equation shows the three cone responses on the left, which are computed as the dot product of the von Kries matrix (D), shown as the 3×3 matrix in (7.3), which are multiplied by the tristimulus values X, Y, Z for the appropriate white point.

$$\begin{bmatrix} \rho \\ \gamma \\ \beta \end{bmatrix} = \begin{bmatrix} 0.40024 & 0.70760 & -0.08081 \\ -0.22630 & 1.16532 & 0.04570 \\ 0 & 0 & 0.91822 \end{bmatrix} \begin{bmatrix} X \\ Y \\ Z \end{bmatrix} \qquad (7.3)$$

Having computed rho (ρ), gamma (γ), and beta (β) for each white point, the ratios of these values can be applied as shown in Eq. 7.4 to complete the transform. As shown, this transform begins by computing the inverse of the von Kries matrix (D), multiplying it by a diagonal matrix containing ratios of the cone responses for each white point, multiplying the result by the matrix D and finally multiplying the result by the tristimulus values in the image. This transformation provides values which

7.1 RGB Display Rendering

when rendered appropriately would create a metameric match between the original image displayed at 6500 K and the modified image displayed at 9300 K, where the user is adapted to the respective color temperatures. Although this process appears complex, each of the matrices on the right can be cascaded into a single matrix, significantly simplifying this process.

$$\begin{bmatrix} X_{9300} \\ Y_{9300} \\ Z_{0300} \end{bmatrix} = [D]^{-1} \begin{bmatrix} \rho_{9300}/\rho_{D65} & 0 & 0 \\ 0 & \gamma_{9300}/\gamma_{D65} & 0 \\ 0 & 0 & \beta_{9300}/\beta_{D65} \end{bmatrix} [D] \begin{bmatrix} X_{D65} \\ Y_{D65} \\ Z_{D65} \end{bmatrix} \quad (7.4)$$

In this particular example, the final correction matrix is shown in Eq. 7.5.

$$\begin{bmatrix} X_{9300} \\ Y_{9300} \\ Z_{0300} \end{bmatrix} = \begin{bmatrix} 0.9848 & -0.0521 & 0.0634 \\ -0.0057 & 1.0042 & 0.0012 \\ 0 & 0 & 1.2987 \end{bmatrix} \begin{bmatrix} X_{D65} \\ Y_{D65} \\ Z_{D65} \end{bmatrix} \quad (7.5)$$

In summary, a matrix is derived using the von Kries matrix to adjust the image so that it will be perceived similarly when presented on a display having a white point of 9300 K as it would have been perceived when captured under D65 illumination. The von Kries transform can be applied in this step. However, other transforms can be applied to perform a similar adjustment. Some of these transforms, for example the Bradford transform [16], are very similar to this transform while others are nonlinear and provide a more complex adjustment.

7.1.4 Target Display Definition

To progress beyond this point, we need to understand the display on which we want to display the image. It is particularly important to understand the primaries and the code value to luminance transform of the display as we define the performance of our target display.

> Beside the display white point, the chromaticity coordinates of a display's primaries and the luminance response of a display often differ from those assumed within any image encoding format. To obtain a metameric match to the original scene, each of these factors must be considered.

In this chapter, we will use a target display based on an addressable array of inorganic LEDs. Like many displays, the current array of iLEDs will be driven by controlling the flow of voltage to the iLEDs. As a result, when we measure the output

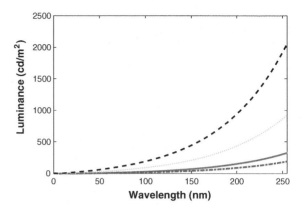

Fig. 7.2 iLED luminance output depicted as a function of code value and LED color

of the iLEDs as a function of code value we obtain curves such as those shown in Fig. 7.2. Note that in this example, the array of iLEDs include red, green, blue, and white iLEDs. In our initial discussion, only the output of the red, green, and blue iLEDs will be applied. The output of the white iLEDs will enter into later examples within this chapter.

As is common, the luminance of the blue iLED is the lowest, followed by red. The green iLED is highest in luminance among the three primaries due to the eye's sensitivity. As shown, the white iLED is higher in luminance than any of the three color channels, partially due to the yellow peak provided by this primary, which is near the peak sensitivity of the human eye. Although perhaps not evident from this plot, the luminance of the white iLED is actually higher than the luminance of the sum of the luminance for the red, green, and blue iLEDs. In fact, the luminance of the white iLED is about 1.4 times the luminance of the sum of the luminance provided by the red, green, and blue iLEDs. This is not atypical. As we discussed earlier, blue iLEDs are particularly power efficient and when these iLEDs are coated with an efficient phosphor or other color conversion medium, the luminance output from these white iLEDs can often easily exceed the luminance output of individual red, green, and blue iLEDs for an equivalent input power. The efficiency of white iLEDs will enter into our discussion later.

For the particular device in this example, the chromaticity coordinates of each iLED were stable once the luminance was high enough to permit a reliable measurement. The 1931 CIE x, y chromaticity coordinates for each primary are shown in Table 7.2. While measurements indicated that there is some variability in the chromaticity coordinates of the iLEDs as a function of code value, this variability is small, usually within 0.002 values, and can be considered to be negligible for the examples in this chapter. If this is not the case and color varies as a function of code value then further correction may be required to reduce this variability. In the examples used, the chromaticity coordinates shown in Table 7.2 will be assumed to represent the chromaticity coordinates of the iLED for the entire tone scale.

7.1 RGB Display Rendering

Table 7.2 Peak luminance value (Y) in cd/m² and 1931 CIE x, y chromaticity coordinates for each primary

Color	Y	x	y
Red	329.7	0.694	0.306
Green	926.3	0.157	0.725
Blue	192.6	0.137	0.051
White	2057.7	0.246	0.210

Having described the basic performance of the target display, we can now turn our attention to rendering the image data onto this display.

7.1.5 RGB Rendering

The goal of the RGB rendering step is to convert the image data to the specific primaries in the display. As noted in Chap. 4, many of the encoding standards assume that the white point of the display has chromaticity coordinates similar to the chromaticity coordinates of standard 6500 K daylight (i.e., D6500). Further these standards assume a specific set of display primaries having RGB primaries with specified chromaticity coordinates. However, comparing the REC709 primaries in Table 4.1 with the display primaries shown in Table 7.2, it is evident that the primaries of the display vary significantly from the chromaticity coordinates that were assumed during image rendering. For example, the assumed green primary has x, y chromaticity coordinates of 0.30, 0.60 as shown in Table 4.1. These are compared with the green primaries for the current display of 0.157, 0.725. Obviously, these numbers are significantly different from one another and rending the images on the display without compensation for these differences will result in distorted colors.

To correctly transform the images to compensate for these differences in primaries, we can use the primary or phosphor matrix to describe the linear combination of light from our primaries by computing the XYZ tristimulus values that a given linear RGB intensity will produce. This can be computed as shown in Eq. 7.6. In this matrix, the columns of the 3 × 3 matrix on the right are filled with the tristimulus values for the display primaries, the RGB matrix represents the linearized code values used to drive the primaries and the XYZ matrix on the right represents the tristimulus values for the desired color. In this matrix, the RGB intensities are scaled such that the input linear RGB intensities (1, 1, 1) typically result in the tristimulus values of the desired display white point. This then defines the maximum luminance, or unit intensity, of each primary.

$$\begin{bmatrix} X_R & Y_R & Z_R \\ X_G & Y_G & Z_G \\ X_B & Y_B & Z_B \end{bmatrix} \begin{bmatrix} R \\ G \\ B \end{bmatrix} = \begin{bmatrix} X \\ Y \\ Z \end{bmatrix} \quad (7.6)$$

Note, however, that to convert the images, we must invert the primary matrix to permit computation of the RGB code values from the desired image XYZ value. We will refer to this inverted matrix as pMat. Further, this inverted matrix must be scaled appropriately to the desired display white point.

To derive the pMat, we first compute a scalar matrix p as shown in Eq. 7.7. As shown, we begin by computing the inverse of the primary chromaticity matrix. This matrix is rotated and multiplied by a vector containing the tristimulus values for the desired display white point (9300 K in our example).

$$p = \left(inv \begin{bmatrix} x_r & y_r & z_r \\ x_g & y_g & z_g \\ x_b & y_b & z_b \end{bmatrix} \right)' \begin{bmatrix} X_{wp} \\ Y_{wp} \\ Z_{wp} \end{bmatrix} \quad (7.7)$$

The resulting vector p then provides a set of scalars we can apply to normalize the final pMat. The pMat is then calculated by rotating the chromaticity matrix and multiplying it by the vector p, as shown in Eq. 7.8. This then provides the pMat, which we can apply to image tristimulus values to determine a set of RGB scalars. This transformation is shown in Eq. 7.9.

$$pMat = \begin{bmatrix} x_r & y_r & z_r \\ x_g & y_g & z_g \\ x_b & y_b & z_b \end{bmatrix}' diag(p) \quad (7.8)$$

$$\begin{bmatrix} R \\ G \\ B \end{bmatrix} = pMat \begin{bmatrix} X_i \\ Y_i \\ Z_i \end{bmatrix} \quad (7.9)$$

At this point, we can begin with the XYZ values for an image from Eq. 7.2, which are rendered to match sRGB primaries with a white point of D6500. These values are then adapted to a white point of 9300 K, using Eq. 7.5. Finally, Eq. 7.9 is applied to render the values to the primaries of the display. At this point, the images are adapted to our desired white point and rendered to take advantage of the primaries of our display. Unfortunately, we have one remaining problem. Reviewing our display characterization, the default white point of our display is not at 9300 K, but slightly magenta of D65 as shown in Fig. 7.3.

It is therefore desirable that we calibrate our display, adjusting driver voltages or look up tables within the display to achieve the desired white point. Although less desirable, we could compensate the image data for this offset of the display's white point. To apply this correction in our image processing chain, we could simply substitute the tristimulus values for the true white point of our display into Eq. 7.7. To determine the relative luminance adjustment for use in calibration, we could substitute the true white point of our display into Eq. 7.7, calculate the alternate pMAT and then substitute the tristimulus values for 9300 K into Eq. 7.9. The resulting RGB

7.1 RGB Display Rendering

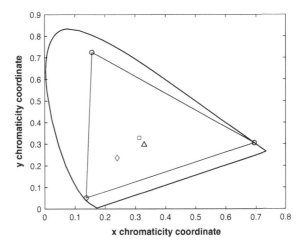

Fig. 7.3 1931 chromaticity diagram depicting display gamut with the D6500 white point (red square), the 9300 K white point (blue diamond), and the default white point of the display (black triangle) in our example

values will provide relative intensities to guide our adjustment of the intensity of the 3 channels, guiding adjustment during display calibration. For example, completing this later calculation provides normalized RGB values of [0.5198 0.5225 1]. The peak luminance values of the display must then be adjusted to have the same ratios as provided. In this example, we might then reduce the intensities of the red and green channels to 171.4 and 484 cd/m^2, respectively. This would result in peak luminance values of 848 cd/m^2 for the display.

7.1.6 Tonescale Rendering

At this point, we have determined code values in a space that is linear with luminance. However, as shown in Fig. 7.2, the response of the display is not linear with luminance. Therefore, we need to compensate for the nonlinear characteristic curves of the display by applying the inverse of the functions shown in Fig. 7.2 to derive final display code values. If we have equations which reasonably represent the luminance response of our display, we can invert these equations to permit us to convert from desired relative luminance values to display code values. An alternate method is the use of a look-up-table. Assuming that we have an accurate calibration curve for our display, we can use this calibration curve to look up the appropriate code values. Appropriate look-up tables for our display are shown graphically in Fig. 7.4.

> A final step within a color rendering path is to convert the image data from a space which is linear with respect to scene luminance to a space that accounts for the nonlinear response of the final display. Many other potential image

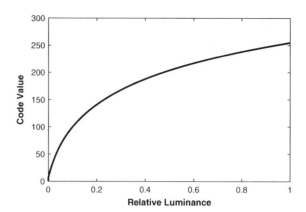

Fig. 7.4 Graphical depiction of lookup tables appropriate for transforming relativeluminance to 8-bit display code values

enhancement algorithms are often employed prior to this step, taking advantage of linear encoding.

7.1.7 Summary

Throughout this section, we have discussed the basic steps necessary to accurately render RGB code values stored in a standard color space to a display at a different white point. Although these steps might seem complex, in practice the matrices in Eqs. 7.2, 7.5, and 7.9 are the only matrices applied to the data. Further, these three matrices can be concatenated into a single matrix. Therefore, the entire image processing chain might require linearizing the image data from the image, applying a single 3×3 matrix, and then adjusting the resulting RGB relative luminance values for the nonlinear response of the display by performing a look-up from relative luminance to display code values. Therefore, the steps performed on the image data can be relatively concise.

Notice that the entire color transformation from one set of primaries to a second set is accomplished with the pMat. This approach requires an understanding of Eq. 7.6 and the ability to invert this matrix to determine the pMat. As we will see in the subsequent section, it is this step that we need to alter when rendering images to multi-primary displays. Further, the inversion of the matrix to determine the pMat must be modified significantly to permit rendering to multi-primary displays.

7.2 Color Rendering for Multi-primary Displays

To begin our discussion of rendering for multi-primary displays, we must first acknowledge that there are deficiencies in the process we have just described which prevent us from applying this process to multi-primary displays. Specifically, as we described Eq. 7.6, we acknowledged that we needed to invert the matrix of tristimulus values on the left side of this equation. In multi-primary displays, we can form an equation similar to the one shown in Fig. 7.6. The resulting equation for a four-color display, where the fourth color is represented by W, is shown in Eq. 7.10.

$$\begin{bmatrix} X_R & Y_R & Z_R \\ X_G & Y_G & Z_G \\ X_B & Y_B & Z_B \\ X_W & Y_W & Z_W \end{bmatrix} \begin{bmatrix} R \\ G \\ B \\ W \end{bmatrix} = \begin{bmatrix} X_i \\ Y_i \\ Z_i \end{bmatrix} \qquad (7.10)$$

As we can see the 3×4 matrix of tristimulus values on the left is not square and therefore the inversion of this matrix is not a valid mathematical operation. Further, we can see that for any image, we have three known values (X_i, Y_i, Z_i) and four unknowns, specifically R, G, B, and W intensity values. Therefore, it is not possible to solve this equation given the current information.

To develop a solution to this problem, we are going to begin with a set of simplifying assumptions. We will then relax these assumptions, one at a time to develop a more general solution to this problem. Much of this solution has been discussed previously [11].

7.2.1 Simplifying Assumptions

In this section, we will begin by making a number of simplifying assumptions. We will then derive a solution with these simplifying assumptions in place. Once we derive a solution, we will then relax one or more of these assumptions, expanding our solution to the increased solution space. We will continue this process until we have relaxed each of these assumptions and have a relatively general solution.

It should be acknowledged that depending upon the overall goal for using a multi-primary display and the characteristics of the display technology to be applied, modifications to this approach may well be desirable. However, the current approach provides a general framework with free parameters which can be adjusted to provide a relatively broad range of solutions.

To develop a solution, the simplifying assumptions we will begin with are the following:

(1) The color of our white emitting element matches our aim display white point (e.g., chromaticity coordinates equivalent to the chromaticity coordinates of

9300 K). This color will be consistent throughout the entire luminance range of the additional light-emitting element.
(2) The maximum luminance of the additional light-emitting element will be equivalent to the sum of the luminance output of the red, green, and blue light-emitting elements.
(3) Our additional light emitting element will output white light in response to intensity W.
(4) Our multi-primary display will have exactly one additional colored light emitting element, in addition to the RGB light-emitting elements.
(5) The display will be calibrated such that the chromaticity coordinates of all colors are constant throughout the entire luminance range of the display.
(6) The goal of the algorithm is to maintain color accuracy. That is, the display should be able to produce a metameric match to a comparable display having only RGB light-emitting elements, where the additional light-emitting element is driven in response to an image signal with the additional primary.
(7) The white point of the display will be held constant, regardless of the input image. This is a traditional assumption in RGB displays. That is a diffuse white should be rendered at a known triad of code values (e.g., [235; 235; 235]) and the neutral scale of the display will be rendered proportionally to white. Therefore, the luminance of a white displayed in an image will be the same under all conditions.

7.2.2 The Color of W Is the Display White Point

Additional white subpixels have been widely discussed for use in both OLED and liquid crystal displays [1, 2, 6, 7, 17, 18, 21]. Each of these approaches often employ a custom method of image processing, which often introduces substantial color error during the rendering process. However, in our discussion, we wish to understand how to perform the transformation to provide a metameric match between a typical RGB and RGBW display, according to our assumptions.

With this set of assumptions in place, when we apply the image processing path that we described through Eq. 7.9 in Sect. 7.1.5, we are able to create RGB intensities where the sum of the luminance created by red, green, and blue light emitting elements to an image signal having equal RGB intensities will be equal to the luminance of the W light-emitting element having a W intensity equal to the RGB intensity. Further, the color of light produced by the sum of the red, green, and blue light emitting element is now equal to the color of light produced by the white light emitting element. In our example, we can create light having chromaticity coordinates equal to the chromaticity coordinates of 9300 K white light using either equal intensities of RGB or the same intensity of W. Under these assumptions, we can decide to form any neutral or white luminance using either light from an equal combination of red, green, and blue light-emitting elements or the W light-emitting element.

7.2 Color Rendering for Multi-primary Displays

We can express the transfer of intensity from the RGB intensities to the W intensity as shown in Eqs. 7.11 and 7.12. As shown in these equations, we must decide upon the proportion of the neutral luminance we want to move from the RGB channels to the W channel. We will call this proportion the White Mixing Ratio (WMR). This value provides a fourth known value, in our process, now providing four knowns and four unknowns and enabling the transformation from RGB to RGBW.

As shown in Eq. 7.11, the W intensity is determined by calculating the minimum of the RGB intensity values (i.e., the neutral portion of the pixel signal) and multiplying this minimum by the WMR. This calculation assumes that a proportion of the neutral luminance will be produced by the W channel. We must then remove this luminance from the RGB channels. Therefore, as shown in Eq. 7.12, modified RGB intensity values (R', G', B') are calculated by subtracting the resulting W intensity value from each of the RGB intensity values.

$$W = WMR * \min\left(\begin{bmatrix} R \\ G \\ B \end{bmatrix}\right) \tag{7.11}$$

$$\begin{bmatrix} R' \\ G' \\ B' \end{bmatrix} = \begin{bmatrix} R \\ G \\ B \end{bmatrix} - W \tag{7.12}$$

Because of the earlier assumptions, these two manipulations produce a set of R'G'B'W intensity values which produce light having the same luminance and chromaticity coordinates as the original RGB intensity values. That is, these manipulations create a metameric match between the RGB only solution and the R'G'B'W solution. One might then ask, "how does one select an appropriate WMR?" The answer is, "it depends upon your goals for utilizing a multi-primary display". For example, in the display of our example, for a given luminance, the W light-emitting element will actually require less power than any of the RGB light-emitting elements to create an equal amount of luminance. Since you are actually replacing the sum of the light from the RGB light-emitting elements with the W light-emitting element, we will save significant amounts of power by creating white light using the W light-emitting element instead of summing the light from the RGB light-emitting elements. If our goal is to save as much power as possible, we might employ a WMR equal to 1, moving as much of the neutral luminance as possible from the RGB light-emitting elements to the W light-emitting element, while maintaining the color information in the image.

> The transform from a 3 color signal to a 3 + n color signal requires the definition of n free parameters. These parameters are selected based upon the goals of the multi-primary display designer. They may be selected to enable power reduc-

tion, but alternately might be selected to enable gamut expansion, improved spatial resolution, or other goals.

7.2.3 Relaxing the White Assumption

In this section, we begin by relaxing assumption number 1. This assumption stated, "the color of our white emitting element matches our aim display white point (e.g., chromaticity coordinates equivalent to the chromaticity coordinates of 9300 K)." We will continue to assume that our additional light-emitting element emits light which is inside the gamut formed by the chromaticity coordinates of our red, green, and blue light-emitting elements. However, the chromaticity coordinates of our W light-emitting element does not necessarily match the white point of our display. Instead the color of the W subpixel may be biased towards one or two of the color primaries with respect to the desired display white point.

Using our iLED display as an example, this condition is illustrated in Fig. 7.5. We see that the color of our white subpixel is biased towards the blue and red color primaries with respect to our aim white point of 9300 K. Under these conditions, if we apply the algorithm discussed in Sect. 7.2.1, the simple expressions embodied in Eqs. 7.11 and 7.12 will introduce color error as we will replace light having chromaticity coordinates which correspond to the chromaticity coordinates of 9300 K white light with blue-magenta light. Therefore, we will not achieve a metameric match if we apply the algorithm provided in the previous section. Instead the neutral portions of the image will appear blue-magenta in color.

To correct for the discrepancy between the color of the display white point and the color of the W primary, we must account for the color of the W primary when transferring luminance from the RGB light-emitting elements to the W light-emitting element. We can graphically understand the desired correction using Fig. 7.5. Once again, this figure shows the white point of the display, and the chromaticity coordinates of each of the four light-emitting elements in our display. Also shown is the gamut of the display, represented by the triangle formed by the RGB primaries. Finally, three additional triangles or subgamuts are depicted where each subgamut is formed by a combination of two of the primaries and the white point of the display. As shown, the chromaticity coordinates of our W primary resides within the subgamut formed from the blue, red and the white point of our display. Therefore, graphically, we can imagine applying a set of transformations as shown in Fig. 7.6. As shown in this figure, we transform our RGB intensity values corresponding to the white point of the display to the color of the W primary (Wcc). This is exemplified by the matrix operation shown in Eq. 7.13.

7.2 Color Rendering for Multi-primary Displays

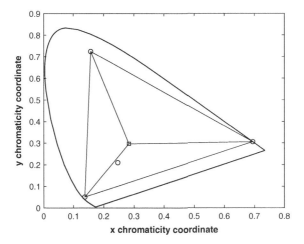

Fig. 7.5 Chromaticity diagram depicting chromaticity coordinates for RGB primaries (indicated by circles at the apexes of the triangular gamut), the display white point (shown as the square connecting the sub-gamuts), and chromaticity coordinates for the W primary (shown as the circle inside the display gamut)

Fig. 7.6 Depiction of image processing path for white point correction

$$\begin{bmatrix} R_n \\ G_n \\ B_n \end{bmatrix} = \begin{bmatrix} W_R & 0 & 0 \\ 0 & W_G & 0 \\ 0 & 0 & W_B \end{bmatrix}^{-1} \begin{bmatrix} R \\ G \\ B \end{bmatrix} \quad (7.13)$$

We then compute the minimum of our normalized RGB values (RnGnBn). This signal (M) is multiplied by our white mixing ratio (WMR) to provide our W intensity value. This W intensity value is then subtracted from each of the normalized RGB values (R_n, G_n, B_n). Finally, the signal resulting from this subtraction is normalized back to the white point of the display, resulting in modified RGB intensity values (R', G', B'). Applying a WMR of 1 as much of the W energy as possible is removed from the color signal. Once again, in our given display configuration, this would transform much of the RGB signal to the W signal as is possible. Given that our W emitter is much more energy efficient than the RGB light-emitting elements, this transform would succeed in significantly reducing the power consumption of the display. Smaller WMR values permit a portion of the W energy to be displayed by the RGB primaries.

These same values can also be used to normalize the final signal to the white point of the display. This transformation is shown in Eq. 7.14. It should be noted that in this process, the luminance of the W primary is not the same as the luminance of the sum of the R, G, and B primaries. Instead, the luminance of the W primary is adjusted such that the W primary and the appropriate luminance values from the red and green primaries form a white point luminance equal to the white point luminance that is created when applying only the RGB values.

$$\begin{bmatrix} R' \\ G' \\ B' \end{bmatrix} = \begin{bmatrix} W_R & 0 & 0 \\ 0 & W_G & 0 \\ 0 & 0 & W_B \end{bmatrix} \begin{bmatrix} R'_n \\ G'_n \\ B'_n \end{bmatrix} \tag{7.14}$$

This algorithm can be applied to any multi-primary display which has a primary in addition to the RGB primaries which is within the gamut formed by the RGB primaries. However, if one is interested in power savings, it is usually desirable for the color of the additional in-gamut primary to be near the white point of the display, as the typical image distribution contains a lot of pixels which have a color very near the white point of the display. In our example, the large percentage of near neutral pixels will be formed from light which is emitted by a combination of the W, green, and red light-emitting elements. While the luminance contribution of the green and red light-emitting elements will be small, all three light-emitting elements will be active when forming a large portion of the colors in the display and the power savings will likely be less than it would be if the color of the W light-emitting element matched the color of the display white point. Of course, if one desires to use the additional primary for reasons other than maximum power savings, placing the additional emitter at locations inside the gamut but not at the white point of the display can have significant advantages as well.

7.2.4 Removing the White Assumption

Thus far, we have assumed that the additional primary is inside the gamut of the display. However, many multi-primary displays have been discussed where the chromaticity coordinates of one or more additional primaries are outside the gamut boundary formed by the RGB primaries [4, 12, 13, 15, 19, 20, 22]. Therefore, it is necessary to be able to relax this assumption as well. The addition of light-emitting elements outside the gamut boundary formed by the typical RGB triad is typically discussed as a method to increase the effective color gamut of a display. The addition of such primaries can have other positive effects. For example, inclusion of a yellow light-emitting element might increase the color gamut of a display as shown in Fig. 7.7. Since the human eye is most sensitive to light near 560 nm, which is usually classified as yellow, the efficacy of the yellow light-emitting element can be greater than that of red, green, or blue light-emitting elements.

7.2 Color Rendering for Multi-primary Displays

If the new primary is outside the RGB gamut, the above discussed method can still be applied. However, a few, relatively minor modifications are required. As before, we will calculate the emitter-equivalent RGB primaries. For consistency, we will continue to refer to these values as W_R, W_G, W_B even though the color of the additional emitter is likely to be yellow or cyan rather than white. It is then important to scale these values such that the maximum of the absolute value of these three values is unity. Note that the absolute value is necessary as at least one of the primaries is likely to be negative. For example, if the additional emitter is a yellow emitter, then W_B will need to be negative. Finally, one additional change is required, specifically when computing the minimum, it is necessary to compute the minimum of only the positive $R_n G_n B_n$ values.

Through the application of these relatively simple changes to the algorithm discussed thus far, a metameric match can be formed between a display having an additional primary outside the gamut boundary of the RGB primaries and the original RGB display. Thus, the original assumption of an additional primary which emits "white" light can also be relaxed using this same image processing path.

7.2.5 Assuming More Than One Additional Light-Emitting Element

It is also possible to remove the fourth assumption, permitting the application of more than one additional light-emitting element by applying this same transformation process. This can be accomplished by simply applying the process depicted in Fig. 7.6 multiple times. With each iteration, the primaries with the largest values can be identified and applied as if they are the RGB values, transferring energy to an

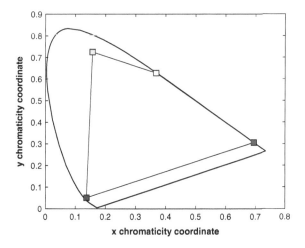

Fig. 7.7 Chromaticity diagram illustrating the inclusion of a yellow-light emitting element having a bandwidth representative of a common iLED but a peak emission at 560 nm. Note that the gamut of the display now has 4 vertices and is no longer a triangle

additional primary. To maximize the use of a primary, this primary should be the last one applied within the iterative process.

> Regardless of the color of any additional primary, one can develop many possible solutions which maintain a metameric match to an RGB display with similar RGB primaries. However, the number of potential solutions increases when the color accuracy assumption is relaxed.

7.2.6 Relaxing the Color Accuracy Assumption

Assumption number 6 is stated as follows: "The goal of the algorithm is to maintain color accuracy. That is, the display should be able to produce a metameric match in response to an image signal with the additional primary as it would if the display had only RGB light-emitting elements." This is an assumption which can be highly desirable. However, flexibility can be achieved by removing or relaxing this assumption. In fact, arguably the most commercially successful class of multi-primary display is the class of projectors utilizing Texas Instrument's Deformable Mirror Device (DMD). These projectors utilized a white light source, which was reflected from a chip containing a large array of mirrors which modulated the light. The light intended for viewing was projected through a color filter wheel, which contained four differently-colored segments, including a red, green, and a blue segment to create primary colors. The color filter wheel also contained a transparent segment, resulting in white light. The algorithm employed within this projector generally performed a process very similar to the process shown in Eqs. 7.11 and 7.12. However, the algorithm did not subtract the neutral component from the RGB components as indicated in Eq. 7.12 [14]. By eliminating this subtraction, additional white light was added to every pixel, substantially desaturating the resulting image. However, the resulting algorithm increased the luminance of the projected image. While the use of this algorithm distorted the color information within displayed images, it nevertheless, led to an overall enhancement in perceived image quality for the vast majority of images within this particular projection application.

> The addition of white luminance in field-sequential color displays, such as Texas Instrument's DMD projectors, not only increased the effective luminance of the display but reduced an imaging artifact referred to as color-breakup. Color breakup occurred in these displays as only one color of light was displayed at any moment in time and when the user made fast eye movements across the displayed image, they perceived color fringing around the edges of

7.2 Color Rendering for Multi-primary Displays

objects. This color fringing was referred to color breakup. The addition of a white image in the projector reduced the visibility of this artifact. This illustrates a point that past, successful commercial application of multi-primary displays has resulted from a solution which overcomes one or more technology or manufacturing limitations in the underlying display technology.

As it is often desirable to introduce at least some color error to increase the effective luminance of the display, it is important to consider algorithms capable of providing an increase in luminance while effectively minimizing the color error.

Recalling the discussion of color saturation and intensity at the end of Chap. 3, we showed in Fig. 3.17 that the probability of observing a highly saturated pixel was small but even this small probability decreases as the relative luminance of a pixel increases. Therefore, the probability of observing a very high luminance and saturated color is low within natural images. While these colors exist in graphical images, the color rarely has meaning and therefore accurate rendering of these colors is often not essential. In most multi-primary displays, distorting these colors often provides an opportunity to increase the luminance of the display, providing advantages similar to those observed for the DMD. However, alternate approaches can achieve increases in luminance with much less color distortion. One approach to this is distort only the high saturation, high luminance portion of the image [10].

Two basic approaches can be applied to achieve the color distortion. The first of these include RGB limiting, in which the linearity of the RGB intensities are altered, decreasing the slope of the linear RGB intensity values above some threshold. The result of this manipulation is to maintain calculated color saturation but to darken the infrequently-occurring bright saturated colors. The second is the addition of extra neutral luminance to the bright, highly saturated colors, decreasing the saturation of these infrequently-occurring bright, saturated colors.

In RGB limiting, a scale factor is computed for any primary (P). This approach is applied after the algorithm depicted in Fig. 7.6 so that the primaries represent predominantly saturated content. These primaries will typically include R, G, and B but if the additional primary includes other non-white or gamut-expanding primaries, a similar approach may also be applied to these values. In this approach, a threshold (thresh) is determined. Values below this threshold will not be modified by the algorithm, while primary intensities above this threshold will be decreased. P_{lim} indicates the maximum intensity for the primary and m represents the slope of the function between the threshold and the limit. This scale value is a value between 0 and 1 and is multiplied with the modified RGB intensity values (R', G', B') from Fig. 7.6 to obtain the scaled primaries (R'_L, G'_L, B'_L).

$$S_P = \begin{cases} 1, & if\ P \leq P_{thres} \\ \frac{m(P-1)+P_{lim}}{R}, & if\ P_{thresh} < P \end{cases} \quad (7.15)$$

Having performed this step, we must then decide the proportion of white luminance to add to each subpixel. Note that the quantity 1-S_p represents the proportional decrease in intensity provided by this algorithm. Multiplying these values for each primary by the proportion of the white point luminance provided by each primary (L_R, L_G, L_B), provides a proportion of the luminance that was removed in the prior step. A portion of this luminance (W_p) can then be added back to each limited pixel through the application of Eq. 7.16.

$$W_{add} = W_{rep} \sum \left(\begin{bmatrix} 1 - S_R \\ 1 - S_G \\ 1 - B \end{bmatrix} \begin{bmatrix} L_R & L_G & L_B \end{bmatrix} \right) \quad (7.16)$$

Note that through this algorithm we have limited the maximum primary luminance value that can be achieved. In an example system, where the maximum achievable white point intensity that can be achieved using the W primary is greater than 1 (e.g., 1.2), we can adjust the white point of the display to an intensity greater than can be achieved using only the RGB primaries. For example, the intensity can be increased to 1.2. Under these conditions, a P_{lim} equal to the inverse of 1.2 can be applied. This approach permits the luminance of the white point to be increased to a value that is greater than can be achieved by applying the RGB primaries while maintaining control of the loss of saturation and luminance for any high luminance, highly saturated pixels within an image without affecting the color saturation or relative luminance of less saturated pixels or dim, saturated pixels. Further, the metameric match is only sacrificed for these high luminance, high saturation pixels. Note that this manipulation also relaxes assumption 2, therefore, the maximum luminance of the additional primary is now greater than the sum of the luminance output of the RGB primaries.

7.2.7 Removing the Stationary White Point Assumption

The final assumption to be relaxed is assumption 7. This assumption proposes that the white point of the display will be held constant, regardless of the input image. This assumption is typical in all traditional RGB displays. The display is only capable of presenting image information as bright as the sum of the RGB primaries and there is little reason to modify this behavior. Perhaps the only exception to this rule are portable displays which are used under a large range of ambient illumination conditions. These displays cannot maximize the luminance of the display under all illumination conditions as this operating condition would provide an unacceptable drain on the batteries in these portable devices. Instead, the peak luminance of the display is adjusted based upon the illumination level, providing high luminance under high illumination conditions and lower luminance when the displayed is viewed under low illumination conditions. Thus these displays attempt to optimize the tradeoff

7.2 Color Rendering for Multi-primary Displays

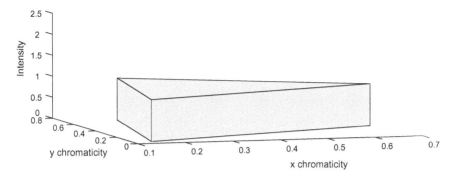

Fig. 7.8 Illustration of a unit intensity color space. The shaded area represents the range of possible chromaticity and relative luminance values

between the rate at which power is drawn from the batteries and visibility of the display.

In certain multi-primary displays, including most RGBW displays, providing dynamic adjustment of the luminance of the display can permit the designer to trade color error for peak display luminance. In these displays, it is common that the white point luminance of the display can easily exceed 2 times the sum of the luminance from the red, green, and blue primaries. In our example, we can activate the RGB primaries to achieve white and then if our W primary is capable of outputting 1.2 times the luminance of the RGB primaries, we can activate the W primary to achieve 2.2 times the luminance output by the RGB primaries. In displays with color filters, such as typical RGBW OLED displays, the W light-emitting element alone can often output a luminance that is more than 3 times higher than the luminance output by the sum of the RGB light-emitting elements. Therefore, these displays could output a white luminance value that is more than 4 times the luminance output by the combination of the RGB primaries. In these displays, images which contain only desaturated colors or only desaturated high luminance colors can be rendered to provide white point luminance values that are several times the white luminance which can be produced by the combination of the RGB primaries. However, the luminance of the display will have to be reduced substantially when images are presented which contain a lot of high luminance, highly saturated colors or else these colors will need to be presented with substantial color error.

To illustrate this point, we can visualize the relative intensity space of a display. Assuming that we are constrained to intensities which can be formed by RGB primaries and the W primary is defined to have an intensity of 1 at the sum of the RGB primaries, we might envision the intensity space of the display as shown in Fig. 7.8. As shown in this figure, all colors are constrained to an intensity no greater than 1, thus forming a prism with the edges of the prism defined by the range of possible RGB intensities and the center of the prism being represented by either the sum of the RGB intensities or a W intensity of 1. As shown, all colors can be rendered within this space.

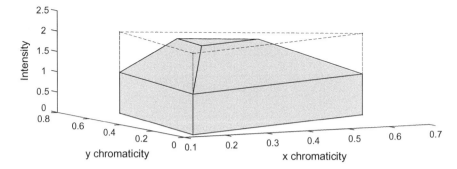

Fig. 7.9 Illustration of an intensity space representing colors that might occur in an image which has been scaled to a maximum intensity of 2, indicated by the wire frame, and physically realizable colors, indicated by the shaded region

Now let's assume that we wish to render colors in an RGBW display to permit an intensity range of 2. All colors having a maximum intensity less than or equal to an intensity of 1 can be rendered as shown in Fig. 7.8. However, as we are expanding the intensity range of the image to 2, we would wish to be able to render all images within the wireframe shown in Fig. 7.9. However, intensity values greater than 1 must involve W intensity values greater than 1 and RGB intensities greater than 1 are typically not available. Therefore, colors having a maximum intensity greater than 1 will be formed from some RGB intensity value of 1 or less and some additional intensity from the W light-emitting element. Therefore, the range of intensities which are physically realizable might be represented by the shaded region of Fig. 7.9. Any colors in an input image which lie outside the shaded region but inside the wire frame are not physically realizable but might be present, although with a low probability, in any input image. When these colors are present, they may be rendered in one of two ways. First, color limiting or white replacement algorithms, such as the ones discussed in the previous section could be employed to map the colors from the wireframe into the shaded region. Another approach is to reduce the maximum intensity until the colors within the image fall within the shaded region. In this second approach, the maximum intensity, and therefore the peak white luminance, of the image can be adjusted dynamically to control the magnitude of any color error.

To provide a high quality display with adjustable intensity range, and therefore adjustable white point luminance, it is important to constrain the changes in white point luminance which can occur. These luminance changes can take one of two forms. First, large, instantaneous changes in white point luminance are possible anytime that a scene change occurs in a video image. Whenever a large change in scene content occurs, the user does not have a reference against which to judge the luminance of the new video scene. As a result, they are typically unable to determine if a change in luminance occurs at this moment in time. Therefore, if a scene containing a large amount of bright, saturated content is replaced with a scene having mostly neutral content, the maximum intensity range and the white point luminance can be

7.2 Color Rendering for Multi-primary Displays

increased substantially, doubling or even tripling the white point luminance without the user being aware of this change. The inverse has a similar effect. That is, the relative intensity range, and therefore, the white point luminance, can be reduced to a half or a third of the original value at a scene change without the user being aware of the change in luminance. However, if there is not a significant change in scene content, the user is able to compare the luminance of one image within a scene with the previous image. Under this condition, the user will be able to detect changes in luminance as small as a few percent per minute. Therefore, the luminance of the display must be adjusted very slowly to prevent the user from detecting unexpected changes in luminance, which would be perceived as an image artifact. As long as these restrictions on luminance changes are observed, the luminance of the display can be dynamically adjusted to trade peak white display luminance for color saturation depending upon scene content [9].

An algorithm for achieving these goals is depicted in Fig. 7.10. In this algorithm, the maximum intensity is selected ahead of time and used to display the image. As the video data is received, it is converted to linear RGB intensity for the proper white point. This image data is used to calculate image statistics, for example, the mean intensity. These statistics can be applied to determine whether a scene change occurred by comparing these statistics between subsequent frames. These statistics might also include the maximum red, green, and blue intensity. The image can then be converted to an RGBW signal. An RGB limit can then be applied to the data and the RGB will be clipped to a maximum value, converting a portion of the lost luminance to the W channel. Finally the image can be displayed. The amount of the signal which is clipped and the image statistics can be used together to select the adjustment for the subsequent frame. This includes selecting a large or small adjustment, as well as determining whether the intensity should be increased or decreased. Finally, a new maximum intensity is selected and used to process the subsequent video frame as it is received.

While algorithms like the one depicted here do not display images with a constant peak luminance as would a traditional RGB display, they permit the display to present as bright an image as possible without introducing unacceptable color error. For most scenes, the peak white display luminance can be maintained at a level that is much higher than the sum of the RGB subpixels. Therefore, the resulting display is viewed as a brighter, saturated display when compared to an RGB display having RGB elements that produce the same luminance as the RGB elements within the

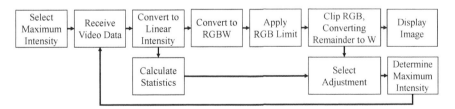

Fig. 7.10 Dynamic intensity adjustment algorithm

comparable RGBW display. Similar algorithms have been discussed for RGBW LCDs (Kwon et al. [5]).

7.2.8 Summary of Multi-primary Image Processing

This section has sought to provide an overview of image processing for multi-primary displays. Many algorithmic approaches are available to this problem. These algorithms are typically designed for use in different display technologies which have different physical limitations. The family of algorithms that are described here were originally developed for use in multi-primary OLED displays. Similar algorithms are currently applied in commercially available OLED products. As the reader may note, there is a bias in the construction of these algorithms to maintain as accurate a metameric match as possible between the multi-primary display and a comparable RGB display. However, this assumption is relaxed to achieve enhancements in overall perceived image quality.

Alternative algorithmic approaches have placed an emphasis on enhancing the brightness of a display or improving the efficacy of a display over the goal of achieving the metameric match. Generally these goals can be achieved through relatively modest changes of the image processing path that has been described, often by adding more luminance from the white light-emitting element than the native algorithms described within the current approach.

7.3 High Dynamic Range

The topic of high dynamic range image capture was discussed earlier. It is important to ask how these images impact display rendering. The answer to this question is likely display technology specific. There are, however, characteristics of a high dynamic range image which are important. First, these images must be presented with high contrast. For most display technologies, this implies that the dark portions of the image must be presented with very low luminance (e.g., substantially less than 1 cd/m^2) and that the highest luminance the display is capable of presenting must be some large factor (e.g., 1,000 to 10,000) times higher than the black level of the display. Finally, there is a requirement that the rendering step must present a diffuse white at some luminance that is substantially greater than black (e.g., 1000 times greater) but at least a few multiples lower than it presents the brightest highlight information. That is, a substantial portion of the luminance range of the display must be reserved for highlight information.

In Chap. 5, we discussed LCD backlights which could be modulated in local areas to permit localized areas in LCDs to achieve black levels as dark as necessary and yet provide localized areas which are bright enough to meet the requirements of an HDR display. One can argue that emissive displays which can achieve a high enough peak luminance meet these requirements by default as the light-emitting elements can typically be turned off, producing black light-emitting elements with 0 cd/m^2 luminance. However, it is necessary in any display technology to be able to reserve enough luminance range above the point at which a diffuse white is rendered to present highlight information with a substantially higher luminance to achieve a high dynamic range display.

As we have noted in Chap. 3, this highlight information is, again, relatively desaturated. Therefore, it is possible in multi-primary displays which may have the ability to present colors with up to 4 times the sum of the luminance of the RGB pixels to present diffuse white as described in the previous section, perhaps rendering it at 2 or 2.5 times the intensity of the RGB pixels, but to retain the additional range of the additional light-emitting element, which can be 4 or 4.5 times the intensity of the RGB pixels in some technologies, for highlight information. Approaches, such as this, which have sought to render even traditional video as HDR by decompressing the tone scale in the highlight regions of the image have been discussed and applied commercially [8].

7.4 Summary and Questions for Reflection

This chapter has discussed an example set of algorithms for rendering images to multi-primary displays. Although many possible algorithmic approaches exist, an approach was introduced which maintains a metameric match to a standard RGB display. The initial application of this algorithmic approach was discussed specifically for RGBW displays, where the chromaticity coordinates of the W light-emitting element is within the gamut formed from the chromaticity coordinates of the RGB light-emitting elements. Modifications of this algorithm were discussed which permits it to be applied to RGBX displays where the chromaticity coordinates of the X light-emitting element are outside the gamut boundary formed from the chromaticity coordinates of the RGB light-emitting elements. Finally, this same algorithmic approach was extended to displays having more than four light-emitting elements.

The discussion then turned to gaining additional dynamic range or higher peak luminance values by permitting mild distortions to the original metameric match. These algorithmic approaches included darkening the RGB light-emitting elements and replacing RGB luminance with other neutral luminance to take advantage of higher luminance addition light-emitting elements. Finally, we discarded an underlying assumption of all RGB rendering approaches, the assumption that the luminance of a white object within a scene should be constant. Instead, we adopted an approach which permitted dynamic adjustment of the white luminance in a display to provide the highest possible luminance with the constraint of maintaining an acceptable colorimetric match to the original RGB display, and even extended this to create a high dynamic range display. While these algorithms were originally developed for RGBW light-emitting displays (OLED or iLED), relatively simple modifications permit their application in LCD or other display technologies.

With this discussion in place, we might consider the following questions for contemplation:

(1) Is it important to reference the rendering path for a multi-primary display to the imaging path for a typical RGB display? If so, why?
(2) Each of these algorithmic approaches required accepting an image encoded in a space where code value is nonlinearly related to scene or display luminance but decoded this image to permit color manipulations to be performed in a linear space. Why is this necessary? What are the implications of this process?
(3) As discussed, commercially successful multi-primary displays have overcome significant technology or manufacturing challenges. What challenges are important to overcome where multi-primary displays may be useful?
(4) There are other algorithms; for example gamut expansion, upscaling, noise reduction; which might be useful in a display but which were not discussed here. How might these algorithms fit within the image processing framework which was discussed?
(5) The text proposed three criteria for a high dynamic range display. Are these three criteria complete? Are all of these criteria necessary?

References

1. Arnold, AD, Castro, PE., Hatwar, TK., Hettel, MV, Kane, PJ, Ludwicki, JE., Matsumoto, S (2005) Full-color AMOLED with RGBW pixel pattern. Journal of the Society for Information Display 13(5):525–535. http://doi.org/10.1889/1.1974009
2. Brown Elliott, CH, Higgins MF, Hwang S, Han S, Botzas A, Hsu B-S, Nishimura S (2008) PenTile RGBW® color processing. In: SID symposium digest of technical papers, vol 39, p 1112. http://doi.org/10.1889/1.3069331
3. Giorgianni E, Madden TE (1998) Digital color management: encoding solutions. Addison-Wesley Longman Inc, Reading, MA
4. Han SH, Kim YH, Yoon JM, Park IC, Jun MC, Jung IJ (2010) Luminance enhancement by four-primary-color (RGBY). In: SID symposium digest of technical papers, vol 41, p 1682. http://doi.org/10.1889/1.3500233
5. Kwon KJ, Kim MB, Heo C, Kim SG, Baek JS, Kim YH (2015) Wide color gamut and high dynamic range displays using RGBW LCDs. Displays 40:9–16. https://doi.org/10.1016/j.displa.2015.05.010
6. Langendijk EHA, Belik O, Budzelaar F, Vossen F (2007) Dynamic wide-color-gamut RGBW display. In: SID symposium digest of technical papers, vol 38, pp 1458–1461. http://doi.org/10.1889/1.2785590
7. Lee B, Park C, Kim S, Kim T, Yang Y, Oh J, Kim C (2003) TFT-LCD with RGBW color system. In: SID symposium digest of technical papers, pp 1212–1215
8. Miller ME, Alessi PJ, Hamer JW, Cok RS (2012) Increasing dynamic range of display output. US Patent Number 8,184,112
9. Miller ME, Basile JM (2010) Luminance and saturation control for RGBW OLED displays. In: SID symposium digest of technical papers, pp 269–272
10. Miller ME, Murdoch MJ (2009) RGB-to-RGBW Conversion With Current Limiting for OLED Displays. J Soc Inform Disp 17(3):195–202. https://doi.org/10.1889/JSID17.3.195
11. Murdoch MJ, Miller ME, Kane PJ (2006) Perfecting the color reproduction of RGBW OLED. In: International congress of imaging science; ISIS 06, pp 448–451. Society for Imaging Science and Technology, Rochester, NY

References

12. Ohsawa K, König F, Yamaguchi M, Ohyama N (2002) Multi-primary display optimized for CIE1931 and CIE1964 color matching functions. In: Proceedings of SPIE 4421, 9th congress of the international colour association, pp 939–942. http://doi.org/10.1117/12.464588
13. Roth S, Weiss N, Chorin MB, David IB, Chen CH (2007) Design and implementation of multiprimary technology. In: SID symposium digest of technical papers, pp 34–37
14. Sampsell JB (1991) White light enhanced color field sequential projection. US Patent Number 5,233,385
15. Teragawa M, Yoshida A, Yoshiyama K, Nakagawa S, Tomizawa K, Yoshida Y (2012) Review paper: multi-primary-color displays: the latest technologies and their benefits. J Soc Inform Disp 20(1):1–11. https://doi.org/10.1889/JSID20.1.1
16. Tooms MS (2016) Colour reproduction in electronic imaging systems: photography, television, cinematography. Wiley, New York, NY
17. Wang L, Tu Y, Chen L, Teunissen K, Heynderickx I (2007) Trade-off between luminance and color in RGBW displays for mobile-phone usage. In: SID Symposium digest of technical papers, pp 1142–1145
18. Wen S (2009) Color gamut and power consumption of a RGBW LCD using RGB LED backlight. In: SID symposium digest of technical papers, pp 1216–1219. http://doi.org/10.1889/1.3256510
19. Xu G, Gille J, Gally B, Tung M-H, Chui C (2008) Optimization of subpixel color tiles for mobile displays. In: SID symposium digest of technical papers vol 39, p 1351. http://doi.org/10.1889/1.3069395
20. Yamaoka R, Sasaki T, Kataishi R, Miyairi N (2015) High-Resolution OLED display with the world's lowest level of power consumption using blue/yellow tandem structure and RGBY subpixels. In: SID symposium digest of technical papers, pp 1027–1030
21. Yoon HJ, Lee JH, Hong KP, Chun JY, Ryu BY, Jun JM, Lee JY (2005) Development of the RGBW TFT-LCD with data rendering innovation matrix (DRIM). In: SID symposium digest of technical papers, vol 36, p 244. http://doi.org/10.1889/1.2036415
22. Yoshida Y, Nakagawa S, Yoshida A, Yoshiyama K, Furukawa H (2011) The luminance resolution characteristics of multi-primary color display. J Soc Inform Disp 19(11):771–780

Chapter 8
Spatial Attributes of Multi-primary Displays

8.1 Replacing RGB Elements in Multi-primary Displays

Remembering back to our discussion of the various display technologies, it is clear that the formation of each light-emitting element within direct view displays is expensive. In liquid crystal or OLED displays, we need to form separate data lines for each column of light-emitting element, TFTs and often capacitors need to be formed for each light-emitting element, and each light-emitting element consumes a portion of the area on the display surface. Forming the additional structures requires technology to form smaller features and the potential for manufacturing defects increases with the addition of each of these structures. Therefore the addition of these data lines, TFTs and capacitors increase the overall cost to manufacture a display.

The need to add more light-emitting elements within the display also reduces the area dedicated to each light-emitting element. For example, we may need to fit four or more light-emitting elements within the same area as the three RGB light-emitting elements. The loss of area for each light-emitting element might translate to a loss of efficiency for an LCD or a reduction in the lifetime for an emissive OLED or iLED display. To avoid these losses in efficiency or lifetime, the designer of a multi-primary display might decide not to make the light-emitting elements smaller. Instead they may elect to increase the size of each full color pixel, reducing the physical resolution of the display. This tradeoff is illustrated in Fig. 8.1. Note the reference arrangement of RGB pixels on the left. Substituting a four-color array of equally-sized light-emitting elements increases the size of the full-color RGBW pixel array as shown in the center of Fig. 8.1. This arrangement results in $3/4^{ths}$ as many full color pixels. The alternate option is to reduce the size of the light-emitting elements to fit the additional light-emitting element in the same area as the original RGB pixel, as shown on the right in Fig. 8.1. Obviously each of these tradeoffs is undesirable.

It is true that one approach or the other might be acceptable if other advantages of the multi-primary display provide a significant enhancement of the underlying technology. In fact, the latest RGBW OLED televisions available from LG Electron-

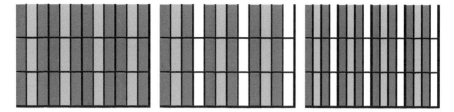

Fig. 8.1 Illustration of resolution tradeoffs when adopting direct view multi-primary displays. Center RGBW arrangement maintains light-emitting element size as original RGB arrangement shown on the left. Right RGBW arrangement reduces light-emitting element size to maintain equal pixel size to RGB arrangement shown on the left

ics incorporate four light-emitting elements in the area usually dedicated for three light-emitting elements because OLED displays formed from a white OLED emission layer with a RGBW color filter array are significantly easier to manufacture than patterned RGB OLED displays. Additionally the use of the RGBW color filter array reduces the power consumption substantially as compared to OLED displays formed with a white emission layer and RGB color filters. However, one should ask "Are there other potential tradeoffs which enable the use of multi-primary displays in direct view displays?" It is this question that we will attempt to address in this chapter.

> The adoption of multi-primary LCDs and LED displays will often require increasing the number of light-emitting elements in a display. This reduces the size of the light-emitting elements and the fill factor of the display. This reduction can reduce image quality, display efficacy and lifetime if adding the additional primaries fails to overcome these limitations.

8.2 Resolving Power of the Human Eye

When discussing the performance of the human eye, we mentioned that the spatial resolution of the eye is significantly higher for the luminance portion of a signal than the chrominance portion of the signal. To illustrate this fact, example luminance and opponent-channel chrominance Contrast Sensitivity Functions (CSFs) are shown in Fig. 8.2.

This figure shows contrast sensitivity (the inverse of the contrast threshold) on the vertical axis. The horizontal axis in this figure represents spatial frequency, specified in cycles per degree of visual angle. These curves are produced by displaying a sine-wave grating, often convolved with a low frequency Gaussian function (typically

8.2 Resolving Power of the Human Eye

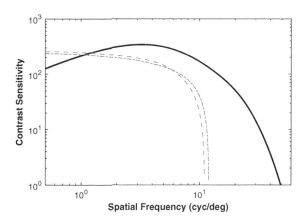

Fig. 8.2 Contrast sensitivity of the human visual system for luminance (black solid line), as well as red-green (red dashed line) and blue-yellow (blue dash-dot line) chrominance signals. Luminance CSF from Barten [2] and chromatic CSFs from Mullen [21]

referred to as a Gabor patch) to an observer. Targets are shown repeatedly with different modulations until the sine wave is found which is just detectable to the human observer. The frequency of the sine wave is used to specify the horizontal axis. For luminance, the vertical axis represents the inverse of the modulation, using the equation shown in Eq. 2.2 back in Chap. 2 which a person needs to detect the presence of the sine wave grating. For the chrominance channel, the vertical axis represents the percent of red or yellow in the mixture of the chromatic grating [21]. As human visual sensitivity is relatively constant for constant visual angles as long as the viewing distance is greater than a half meter, the frequency is then depicted in these units [7]. Note that visual angle refers to the angle subtended by a stimulus or target on the human retina.

As Fig. 8.2 shows, at a given luminance value, the luminance sensitivity of the human eye extends to higher frequencies than either of the opponent-channel chrominance CSFs. That is the luminance sensitivity extends to higher spatial frequencies (i.e., is sensitive to finer spatial patterns) than the chrominance CSFs. Specifically, this figure illustrates that at high contrast, the human eye can detect luminance gratings with a frequency greater than 40 cycles per degree. On the other hand, chrominance gratings become indistinguishable for frequencies just over 10 cycles per degree. It is also interesting that the contrast sensitivity function for the luminance channel is generally a band-pass function, having a peak sensitivity at 2 to 4 cycles per degree of visual angle, while the CSF for the chrominance channel are low pass filters, being equally sensitive to low spatial frequencies before declining for higher spatial frequencies.

All light-emitting elements provide both luminance and chrominance information. This fact complicates the application of known visual phenomena to optimize display pixel patterns.

While the fact that the human eye is less sensitive to high frequency chromatic signals than luminance signals is interesting, it may not be immediately clear how we would apply this information in display design. After all, each and every light-emitting element produces a portion of the luminance signal and generally every light-emitting element will contribute to the chrominance signal. As we discussed the chrominance signals earlier, we mentioned that some models of human vision consider that the retina provides two chrominance signals to the human brain, in addition to a luminance signal. In these models, one of these chrominance signals is composed of a difference between the red-green chrominance signals and the second chrominance signal is composed of a difference between the blue-yellow chrominance signals as they are depicted in Fig. 8.2.

It is also necessary to recognize that each color of light-emitting element generally carries different amounts of luminance information. Generally, red and blue light-emitting elements provide less luminance than the green light-emitting element. As a rule of thumb, when we add an additional light-emitting element, we will add white or yellow light-emitting elements, which provide even higher luminance than the green light-emitting element. However, we might alternately add cyan light-emitting elements, which typically have luminance output between the luminance output of the blue and green light-emitting elements, or magenta light-emitting elements which typically have luminance output between the luminance output of the red and blue light-emitting element.

> Pairs of complimentary colors (red-cyan, blue-yellow, green-magenta) generally each produce a high luminance signal. If we are only interested in rendering luminance with high resolution, we can apply each pair of complimentary light-emitting elements to render the luminance signal in an input image. That is, the luminance of each pixel in an input image can be rendered using a complimentary pair of light-emitting elements. While this concept only applies directly to multi-primary displays having six colors of light-emitting elements, this concept can be approximated in four-color displays by applying green to approximate cyan and white to approximate yellow.

To take advantage of this information, let's start by assuming that we only care about the resolution of the luminance channel. Let's also assume that we are considering adding a yellow light-emitting element. In this case, we can form a color of light that is high in luminance by combining the light output from the red and green light-emitting elements or by combining the light output from the blue and yellow light-emitting elements. As a result, we can form an approximation of the luminance signal through the light output from either pair of light-emitting elements. If we assume that these pairs can, to a first approximation, represent the luminance signal, then we can replace the three RGB light-emitting elements by only two light-emitting elements as shown in the center pane of Fig. 8.2 and still create the same luminance resolution for the display. That is, we can replace the RGB triad in the

8.2 Resolving Power of the Human Eye

original display with either a green-red or a blue-yellow pair of light-emitting elements. Thus, as shown in Fig. 8.2 we might adopt an arrangement of light-emitting elements where we form two of the four light-emitting elements of the multi-primary display within the area of the original RGB triad. A similar effect can be achieved in a RGBW display by forming the second luminance pair from a blue-white pair rather than a blue-yellow pair, as shown in the right panel of Fig. 8.3.

With either arrangement, we can create a high luminance signal from two light-emitting element pairs, providing a potential improvement in spatial resolution. However, we have a significant problem. If we look at the distance between individual light-emitting elements of the same color, we can see that this distance is much larger in our new multi-primary light-emitting elements than in our RGB display. For example, the furthest distance between neighboring green light-emitting elements has nearly doubled for our RGBY or RGBW arrangements than for the original RGB arrangement. Should we attempt to display a flat field of green, we have the potential of seeing bright green light-emitting elements forming stripes with black space between them. This image artifact, often referred to as the "screen-door" effect, can be quite disturbing to the user. As a result, this arrangement of light-emitting elements is less than desirable, unless the display resolution is high enough to eliminate this screen door effect.

It is possible to rearrange our subpixels to decrease this maximum distance and the visibility of this artifact. An example arrangement is shown in Fig. 8.4. Once again, we illustrate a comparison of RGBY in the center and RGBW on the right to a comparable RGB arrangement on the left. Note that the maximum distance between each pair of green light-emitting elements is decreased in this arrangement. This arrangement can be sized so that four differently-sized light-emitting elements fill the same area as the RGB triad as is commonly discussed in the literature [26]. However, it is also possible to size the subpixels as shown in Fig. 8.4. As shown, each column of light-emitting elements now contains two colors of light-emitting elements, which is atypical for traditional direct view displays. In this arrangement, the same maximum distance is shared by all light-emitting elements and this distance is approximately the same as it would be in a comparable RGB display. As a result, this arrangement

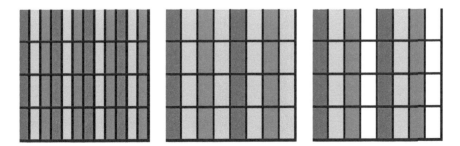

Fig. 8.3 Two light-emitting elements per pixel arrangements as compared to a traditional RGB display which is shown on the left. Note RG and BY (center) or BW (right) pairs of light-emitting elements each produce a high luminance signal with each pair replacing an RGB triplet

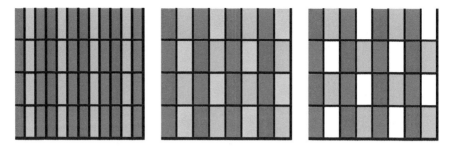

Fig. 8.4 Alternate arrangements of RGBY (center) and RGBW (right) as compared to a traditional RGB arrangement of light-emitting elements (left)

helps to reduce the screen door effect in all colors. This arrangement has another advantage. Specifically every row contains all four colors of light-emitting elements and every pair of columns contain all four colors of light-emitting elements. While other arrangements could be utilized, this arrangement permits one to render a single line of any color using a single row or a pair of columns of light emitting elements.

Not only the number of differently-colored light-emitting elements but the arrangement of the light-emitting elements have a significant influence on the perceived image quality of the final display.

We made the assumption that we could form an approximation to the neutral luminance in a display using red-green, blue-yellow, or blue-white pairs. While each of these pairs of light-emitting elements can provide a high luminance signal, they are not capable of forming truly neutral luminance in all cases. Figure 8.5 illustrates this color mixing phenomena. The left panel of Fig. 8.5 illustrates the RGBY arrangement of light-emitting elements. As we can see the inclusion of a white-light emitting element with a peak wavelength near the peak sensitivity of the human eye (i.e., 560 nm) allows us to expand the gamut of the display slightly from the RGB triangle to the RYGB quadrangle shown in the left panel of Fig. 8.5. Also shown in this figure are two dashed lines connecting the UCS values for the red-green and blue-yellow light-emitting elements. These dashed lines represent the colors of light which can be formed by these pairs of light-emitting elements. While the dashed line between the blue-yellow light-emitting elements is not far from the D65 white point of the display, the dashed line between the red and green light-emitting elements is quite distant from the D65 white point. The implication, of course, is that it will be necessary to add a significant amount of blue energy to the red-green pair if truly neutral luminance is required.

One other point worth making for RGBY displays is that the green light-emitting element selected for a RGB display might not be ideal for a display having RGBY light-emitting elements. As shown in this diagram, the inclusion of the yellow light-

8.2 Resolving Power of the Human Eye

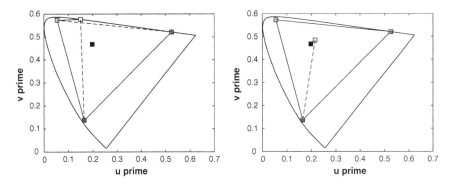

Fig. 8.5 Chromaticity diagrams illustrating that near-neutral luminance can be formed from only pairs of light-emitting elements for displays containing additional yellow (left) or white (right) light-emitting elements

emitting element does little to expand the gamut of the display. However, if the chromaticity values of the green light-emitting element are moved towards lower u' and perhaps lower v' values, the advantage of the yellow light-emitting elements increases and the dashed line connecting the red and green light-emitting elements migrates towards the white point of the display. Therefore, in RGBY displays, shifting the color of the green light-emitting element towards cyan can be beneficial.

The right panel of Fig. 8.5 illustrates a similar arrangement and color mixing for a RGBW arrangement. In this arrangement, we do not have the flexibility to shift the chromaticity coordinates of the green light-emitting element towards cyan as such a manipulation will reduce the purity of our green light. Therefore, in this arrangement, we cannot truly make neutral using only pairs of light-emitting elements. However, we require very little, if any, blue light to create neutral luminance from the white light-emitting element. Therefore, we can use this blue together with red and green light-emitting elements to form neutral light.

One other alteration which can be made is to reduce the number of low luminance subpixels within an arrangement of light-emitting elements. For example, many pixel patterns for multi-primary displays have been suggested which have fewer lower luminance blue and red light-emitting elements than higher luminance green or additional light-emitting elements [3, 20]. While these patterns can be useful as they reduce the number of light-emitting elements per pixel within multi-primary displays, they often suffer artifacts due to the decreased number of light-emitting elements. The use of luminance pairs of light-emitting elements, as discussed in this section, often provides fewer artifacts with similar advantages (i.e., the ability to reduce the number of light-emitting elements per pixel).

> Subsampling the number of low luminance light-emitting elements (i.e., applying a smaller number of blue and potentially red light-emitting elements than

green light-emitting elements) can reduce the overall number of necessary light-emitting elements. However, such patterns are only advantageous when patterns such as those shown in Fig. 8.4 do not quite provide sufficient resolution to replace traditional RGB stripes or higher resolution multi-primary light emitting element arrangements.

8.3 Display Size, Addressability and Viewing Distance Effects

In the previous section, we discussed the resolution of the eye to chrominance and luminance. This discussion may be difficult to interpret. Practically, the implications of this discussion results in three or perhaps four image artifacts which we have to be concerned about avoiding. These can be illustrated by the simple graphic shown in Fig. 8.6. This simple graphic shows transitions from a flat white field, to a gray field, and finally to a red field. In this simple graphic, we expect to see three flat fields separated by clean sharp edges. However, this perception can be easily disrupted.

As we discussed earlier, the first of the disruptions in multi-primary displays is an artifact where one or more fields do not appear uniform. Instead, we see the space between light-emitting elements in the display, creating some form of the screen-door effect. In fields of pure color (e.g., red, green, and blue) this artifact can only be avoided by placing the light-emitting elements close enough together that the human eye cannot resolve the distance between neighboring light-emitting elements of the same color. As a general rule of thumb, this artifact can be avoided as long as the maximum distance between any two light-emitting elements of the same color is separated by a visual angle of 1 min of arc or less [19]. We have more flexibility in neutral colors as it is possible to use most, if not all of the light-emitting elements, to form the neutral color to improve the uniformity of neutral portions of an image.

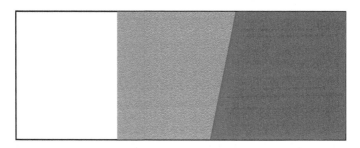

Fig. 8.6 Illustration of a pair of image transitions, relevant to the discussion of image rendering

8.3 Display Size, Addressability and Viewing Distance Effects

As we discuss pixel patterns, we might assume that the designer has significant flexibility to arrange and size the light-emitting elements to avoid effects like this screen-door effect. In fact, a limited number of industry standard display formats with limited addressability place significant constraints on this selection. For example, televisions available today are typically sold as either High Definition TeleVision (HDTV), having 1920 horizontal by 1080 vertical full-colored pixels; as Quad HD, having 3840 horizontal by 2160 vertical full-colored pixels; or 8K displays having 7680 horizontal by 4320 vertical full-colored pixels. Commonly, these displays come in different sizes but most displays have a 16:9 aspect ratio, which defines the ratio of the horizontal display dimension to the vertical display dimension.

While slightly more flexibility exists in the desktop display market, which might include displays having 2560 horizontal by 1440 full colored pixels or a few less common addressable arrays, the resolutions in this market are similar to those used in the television display market. Finally, the portable display market, even cellular telephones, are commonly adopting similar numbers of addressable full-colored pixels within equal areas. However, these devices are often employing displays with longer aspect ratios, increasing the number of full-colored pixels to adjust for the longer relative dimension of the display.

In effect, the number of pixels which can be applied in a display is then set at one of these four values. The size of the displays do vary over quite a large range. However, it is often assumed that the viewing distance of the display generally changes as the size of the display changes. For example, the design viewing distance for the original HDTV specification was assumed to be three times the height of the display. For a typical 32" (0.81 m) HDTV, which has a diagonal of 32", the height of the display is 0.4 m, translating to a viewing distance of 1.2 m. While this is a relatively near viewing distance for a television, the 3 picture height assumption is more realistic for normal viewing of desktop and cellular telephone displays.

Figure 8.7 shows the minimum separation between two light-emitting elements having the same color for both the RGB stripe pattern and the RGBY and RGBW patterns shown in Fig. 8.4. As these particular RGBY and RGBW patterns contain same colored light-emitting elements that are separated by 3 light-emitting elements, each of which is larger than the 2 light-emitting elements that separate similar-colored light-emitting elements in the RGB stripe pattern, the separation between these light-emitting elements is larger. In fact, if we assume a 3 picture height viewing distance and apply the 1 min of arc separation rule, we can see that regardless of the number of addressable light-emitting elements, the RGB stripe pattern always meets the criteria. However, applying this same viewing distance and the 1 min of arc separation rule for the RGBY and RGBW patterns, only the Quad HD and 8K addressable conditions fulfill the criteria. Therefore, these patterns can only be considered for displays having large numbers of addressable pixels, including displays having Quad HD or higher numbers of addressable pixels.

However, this discussion also illustrates the fact that as the resolution of the display is increased, the use of pixel patterns with fewer light-emitting elements per pixel is feasible and using pixel patterns such as those shown in Fig. 8.4 could permit a significant reduction in the number of light-emitting elements in very high reso-

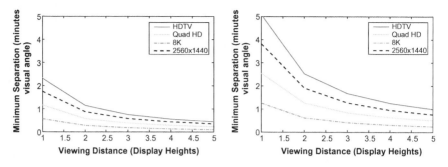

Fig. 8.7 Maximum separation between pairs of same colored light-emitting elements expressed in visual angle. Graph on left illustrates separations for a typical RGB stripe display while the graph on the right illustrates separations for the RGBY and RGBW patterns illustrated in Fig. 8.4

lution displays. In fact, arrangements could be adopted which reduce the number of light-emitting elements compared to traditional four-colored light-emitting element designs by a factor of two and reducing the number of light-emitting elements by one third as compared to traditional RGB stripe pixel patterns.

> The utility of light-emitting element arrangements with fewer than 3 light-emitting elements per pixel of input information to a display is highly dependent upon the addressable resolution and viewing distance of the display. Some arrangements will appear equivalent to RGB stripe patterns at a given viewing distance. However, if the user views the display from a nearer viewing distance, artifacts may become apparent.

8.4 Spatial Signal Processing for Multi-primary Displays

Having addressed the screen-door effect, it is useful to return to our discussion of other image artifacts which can occur in multi-primary displays. The second disruption to our perception of the image shown in Fig. 8.6 is the simple loss of sharpness in the transition between one field and the other. That is, the image is blurred with respect to the original image. This artifact can occur as a function of display configuration if the pixel size of the display is increased, as we demonstrated in the center panel of Fig. 8.1. It can also be produced by image processing, applying multiple columns of pixels to portray this transition.

A third disruption is the appearance or an increase in the visibility of steps along an edge which is not aligned with the light-emitting elements in a display. That is, in the transition between gray and red, we will begin to see the edge as a jagged series

8.4 Spatial Signal Processing for Multi-primary Displays

of stair steps rather than as a smooth line. This artifact can occur due to an increase in pixel size. However, one can imagine that for the red edge, the red subpixels in even the preferred arrangements of light-emitting elements shown in Fig. 8.3 are spaced such that one might see artifacts along the red edge. However, because the eye is not sensitive to high spatial frequency chrominance information, the color of the edge information is not as critical as the luminance information on sharp edges. Therefore, we might apply light-emitting elements other than red to render this edge. However, if we over utilize other colors of light-emitting element, it is possible to see that the edge is comprised of colors other than gray and red. These color artifacts along the edge might be considered either a type of this third disruption or a fourth disruption, if you are counting.

As we discuss rendering images to account for spatial artifacts in multi-primary image displays, this discussion of artifacts illustrates that the image artifacts can be different in areas of an image which contain flat fields of color than areas in an image comprised of edges. In the flat areas, we are generally going to dedicate appropriately colored pixels, using image processing approaches as we described in the previous chapter. However, when edges are present, it can be useful to prioritize applying the light-emitting elements to create smooth spatial structure rather than maintaining color accuracy. A crude representation of the application of this rendering approach to render the transitions in Fig. 8.5 is illustrated in Fig. 8.8. Note that many of the light-emitting elements are enabled, many of them at low luminance, near the illustrated edge regions. Various algorithms have been proposed to attempt to achieve this goal. These algorithms often consist of two different classes.

One class of algorithms purposefully adjusts the mixing ratio used for color rendering in multi-primary displays to purposefully use more of the subpixels in edge regions of an image [23]. This class of algorithm can be employed in addition to a more traditional anti-aliasing or subpixel rendering algorithm. Algorithms such as these have been shown to enhance the perceived quality of pixelated displays, including multi-primary displays [14].

In general, anti-aliasing algorithms recognize that each pixel of data entering the display is a sample which is centered at the center location of the pixel. However, this pixel of data is rendered onto multiple light-emitting elements, many of which

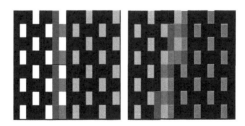

Fig. 8.8 Crude illustration of potential pixel activations used to render the transitions within Fig. 8.5. Note luminance of pixels in edge region has been increased dramatically beyond the level used in practice to improve visibility

are spatially offset in either one or two directions from the center of the pixel of data. Therefore, the correct value for these offset light-emitting elements is not the value input for the pixel but some blend of the value input for this pixel and a neighboring pixel which brackets the location of the light-emitting element. Therefore, if a light-emitting element lies half way between two pixel centers, then its value is likely the average of the values for these two pixels. If it lies at some other point between these two pixels then the value of this pixel can be estimated by calculating its value using the ratio of the difference in the luminance between the two values and the relative difference in spatial distance from the two subpixels. It is this general approach of resampling the image signal based upon the spatial location of the center of the light-emitting elements and the center of the two pixels which is generally referred to as anti-aliasing. This approach can result in color error and other artifacts if the light-emitting elements are large enough that the human eye can detect the color error.

In the RGB literature, this problem was addressed relatively elegantly by Platt [22]. Platt recognized that if filters to accomplish this resampling could be formed which minimized the visibility of error between the image sampled at the location of each element and an ideal unsampled image, then these filters could be applied to provide an optimal rendering. This approach, termed optimal filtering for patterned displays, was adopted for rendering information to RGB striped patterns and was later extended to other, non-striped RGB patterns [12, 30].

This problem becomes a little more complicated for multi-primary displays as the conversion from RGB to RGBX adds another unknown which somewhat confounds the solution. Nevertheless, multiple solutions to this problem have appeared in the literature [8, 9, 17, 18]. Each of these approaches can be used to determine a method of resampling a signal to render an image on a multi-primary display. However, the approaches are specific to the primaries selected, the light-emitting element arrangement, and at times the resolution and viewing distance of the display.

> Although careful design of appropriate rendering algorithms is important to final display quality, the final design is dependent upon the light-emitting element arrangement, the different colored light-emitting elements that are applied, as well as the addressable resolution of the display and the design viewing distance. Each of these factors have a significant effect upon the optimal algorithm.

8.5 Evaluation of Pixel Arrangements

Each of the pixel arrangements we have discussed in this chapter, as well as other available patterns, will have an influence on the perceived image quality of the display.

8.5 Evaluation of Pixel Arrangements

Further, the image rendering algorithms which are present will interact with the selected pixel patterns to influence perceived image quality. Additionally, the relative image quality of these patterns and image rendering algorithms will vary as a function of viewing distance. For example, let us suppose that we were to construct a pair of HDTV resolution televisions having a 32 inch diagonal and view them from 3 picture heights. Let us further suppose that one of these displays had the RGB stripe pattern and the other had one of the two pixel patterns on the right of Fig. 8.4, where there are two light-emitting elements per pixel. Regardless of the image rendering method that we applied, because of the visibility of the screen door effect and other image artifacts, practically everyone would agree that the RGB display was higher in image quality. However, if we were to double the viewing distance or half the pixel size, then the image artifacts would not be visible in the RGBY or RGBW displays. In this case, there would be no perceived difference in image quality and if the RGBW or RGBY displays actually provided higher luminance, then these displays would be judged as higher in image quality than the RGB displays. Therefore, the perceived image quality differences between RGB displays and multi-primary displays are difficult to assess and assessment involves more than simple calculation of contrast or specification of the number of addressable pixels [4].

It is also worth noting that the evaluation we discussed in the previous paragraph is an expensive proposition during display development. To accomplish this assessment, we would need to design and construct prototype displays with each version of pixel pattern to perform this assessment. Without the ability to predict the image quality and the patterns that are likely to be successful, it might be necessary to evaluate several different patterns, making the design, construction and evaluation of displays with many pixel patterns necessary. Obviously, this approach would be overly expensive and likely prohibit the exploration of alternate pixel patterns, making it difficult to replace the standard RGB pixel pattern. Towards this end, a few different approaches have been developed to evaluate and screen various pixel pattern options.

> Construction of prototype displays having various arrangements of light-emitting elements can be very expensive. Therefore, it is important to develop lower cost tools to evaluate and select light-emitting element arrangements and companion rendering algorithms to reduce the risk of investment in multi-primary displays.

8.5.1 Display Simulators

A first approach to evaluating various pixel patterns is to simulate the appearance of various displays. In this approach, a high resolution display simulator is constructed

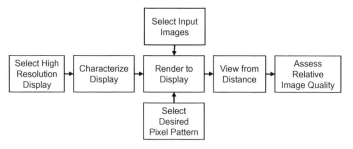

Fig. 8.9 Process applied for simulated pixel pattern assessment

and used to support visual assessment. This method was originally applied by Silverstein and colleagues to permit the assessment of alternate RGB pixel patterns [24] and has been applied more recently in the assessment of alternate multi-primary displays [19]. This involves an approach similar to the one shown in Fig. 8.9. In this approach, a high resolution display is selected to support the display simulator. This display is well characterized and ideally has the ability to create the luminance range of the target display, as well as very dark blacks.

At the heart of the simulator is a suite of software which is able to accept a collection of input images or graphics files and render them using a specified pixel pattern onto the target display. This rendering step is quite different than traditional rendering as it involves the step of rendering an image as though it is being rendered on a display having the desired pixel pattern. To accomplish this, many pixels on the high resolution display are used to represent a pixel in the target display. For example, let's suppose that we wished to render a RGB pixel where there were horizontal gaps between light-emitting elements that were one third as wide as the RGB stripes. To render this pixel pattern, we might render every RGB pixel into a 12 by 12 array of pixels on our high resolution display. Each of the vertical RGB light-emitting elements could be represented by a band of color 3 pixels wide on our high resolution display with a 1 pixel wide black gap between each of the RGB light-emitting elements. Therefore, we see the individual stripes, as well as the area between the stripes. To decide upon the pixel grid to be used, we would have to evaluate all of our potential pixel patterns, determining the minimum number pixels in our display simulator that we could use to form these pixels with some intervening black area between the pixels. We would then need to determine the least common multiple among our pixel patterns and render the pixel patterns to a size equal to our least common multiple. Alternately, we might distort some dimensions slightly to permit the use of as few pixels as possible in our display simulator to represent a pixel in our final display.

For example, let's suppose we wanted to compare the RGB pixel pattern we just discussed to a very similar RGBW pixel having equal dimensions. To render a similar RGBW pixel pattern would require a 16 × 16 grid of pixels in our simulator (3 pixels for each of the four colored stripes) with one intervening black pixel between each group of 3 pixels. We would then need to scale our RGB pixel to a 16 pixel grid,

8.5 Evaluation of Pixel Arrangements

rendering the RGB stripes as 4 pixels wide each, leaving a 1 pixel wide gap between the red-green and green-blue stripes with perhaps a 2 pixel wide gap between the RGB pixels. Such a pattern would create nearly equivalent-sized RGB and RGBW pixels in our display simulator.

Once a pixel arrangement is designed, the images can be rendered onto the display where one pixel of the input image is rendered into each pixel pattern. Therefore, if our simulator requires 16×16 pixels on our high resolution display to represent 1 pixel in our simulated display, we would render each pixel from the image so that the red signal is depicted by all pixels in the red light-emitting element, the green signal is depicted by all pixels in the simulated green light-emitting element, etc. Note that the simulated image processing path will include the color conversion from the RGB signal to the multi-primary display signal and provide whatever method of display rendering which is necessary. At this point, each of our simulated pixels are represented by a grid of pixels on our high resolution display.

Note that at this point, if we viewed our display simulator at a normal viewing distance, we would see representations of these really large pixels, but the image would be difficult to see because the pixels are so large. To truly simulate the entire display, either optics can be applied to minify the image so that pixels subtended the same visual angle as our target display or we could assume that image quality is constant for any visual angle and simply increase our viewing distance to the display such that a pixel on our simulated display subtends the same visual angle as a pixel on the display we wish to construct. At this point, the relative image quality between our example RGB and RGBW pixel representations should approximate the relative image quality of a constructed RGB and RGBW display. It is true, however, that we will likely only be able to display a small portion of an image.

At this point, image quality assessment is possible and any number of subjective image quality assessment techniques might be applied. However, as one wishes to provide a relatively critical assessment of the image quality, a paired comparison method is often applied [10]. In this approach, numerous individuals may be shown each of a number of images rendered with the RGB and RGBW pixel patterns and asked to decide which of the two images appear higher in image quality. Each image may be shown to each individual one or more times. The results of this paired comparison can then be analyzed one person at a time or grouped across individuals and exposed to a scaling method [27]. Such a method will then result in a relative image quality score for each pixel pattern, permitting the image quality of each pixel pattern to be compared in terms of Just-Noticeable-Differences in image quality from a reference (e.g., the RGB pattern). As such, this method can permit one to estimate the relative perceived image quality difference between pixel patterns before investments are made to design or construct a prototype display.

The combination of display simulation with subjective methods and objective image quality metrics provide complimentary technologies which permit the assessment and validation of various arrangements of light-emitting elements in multi-primary displays.

8.5.2 Visible Difference Predictors

An alternate approach to understanding the perceived image quality differences between pixel patterns before constructing a prototype display is to apply Visual Difference Predictors (VDPs). Daly originally discussed the application of VDPs in the context of image compression [6]. In this application an original image was selected. A copy of this image was then compressed. The VDP then permitted one to compare the original and the compressed image to determine if any differences existed between the original and the compressed image. The VDP then mathematically represented the human visual process. This mathematical model of the human visual system was applied to both the original image and the compressed image and the differences between these "mathematically perceived" images were identified as visible differences. In this application, any differences were introduced by the compression algorithm and it was clear that these visible differences corresponded to image degradations which were introduced by the compression algorithm.

Although Daly's model was useful in the assessment of perceived image quality for compression algorithms, which operated predominantly on the luminance information within an image, it did not include any representation of human color vision. A similar but simpler algorithm, called S-CIELab, added spatial filters to the CIE L*a*b* metric which we discussed in Chap. 2 [29]. The general flow of this algorithm is shown in Fig. 8.10. This algorithm provided the function of a VDP, permitting the assessment of regions within an image that had undergone change. This is depicted in Fig. 8.10 by the fact that a similar path representing functions performed by the human visual system is performed on both a reference image and a modified image. Once the data is provided in a space representing what we might perceive, a difference calculation (calculation of ΔE) is performed to determine areas where visible changes occur. Once again, however, this image change was assumed to be a degradation due to the introduction of image artifacts through some image processing step. For example, among the first applications of S-CIELab was to determine the visibility of halftone artifacts introduced during color inkjet printing [28]. The key to this approach is the separation of the image into independent color channels, representing luminance, red-green chrominance, and blue-yellow chrominance, then applying different spatial filters representing human contrast sensitivity to these three channels. Thus spatial or chrominance information that is too high in spatial frequency to be perceived are removed from the resulting images before the difference calculation is performed.

8.5 Evaluation of Pixel Arrangements

This algorithm has then been applied to compare the relative difference in images rendered onto various RGB and multi-primary pixel formats [5]. In this work, images were rendered onto each pixel pattern under consideration, including the traditional RGB stripe pattern. Additionally, the images were rendered onto an ideal pixel pattern, where all three colors were represented in a spatial area covering the entire pixel, as though each color could simultaneously represent all necessary colors. The S-CIELab algorithm was then applied to compare each pixel pattern to the ideal pixel pattern. This research demonstrated that at appropriate viewing distances, the degradation introduced by the multi-primary pixel patterns was no greater than the degradation introduced by the RGB pixel pattern, while artifacts did become more visible (i.e., the ΔE values were larger compared to the ideal pixel) for many of the multi-primary pixel patterns than for the RGB pixel patterns for closer viewing distances. While this result is consistent with results from visual experiments, little further validation of this approach was provided.

Similar studies have been conducted to evaluate subpixel rendering algorithms and different RGB pixel patterns [9]. Once again, realistic RGB pixel arrangements were compared to ideal pixels and differences were calculated. These results have yielded a preference ranking of subpixel rendering algorithms and demonstrated through display simulation and human subject experimentation that the results obtained using the S-CIELab model generally aligned with the results obtained from human judgements of image quality [9]. Based on this one study, it appears that this approach can produce a reasonable assessment of various pixel patterns and the algorithms which are applied to render images onto these pixel patterns.

8.5.3 MTF-Based Evaluation Methods

Because the pixel patterns of interest are color, there is often a desire to apply color metrics to assess the image quality of various pixel patterns. While this tendency is understandable, it may not always be necessary. The fact is, different pixel patterns will produce different luminance patterns and the information capacity of these luminance patterns might be quantified using a luminance-based <u>M</u>odulation <u>T</u>ransfer <u>F</u>unction (MTF) metric. MTF refers to the amount of contrast or modulation which

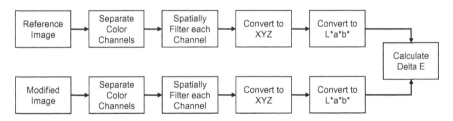

Fig. 8.10 Typical image processing flow applied in S-CIELab

is passed by the system. MTF may be measured by passing a temporally-varying sinusoidal signal with a maximum amplitude into a display and measuring the luminance output across the surface of the display. Generally displays will pass frequencies lower than 0.1 cycles per pixel with near maximum amplitude but the modulation, specified as Michelson contrast, will decrease for higher spatial frequencies. It should be noted that the MTF is a frequency-based metric, much like the contrast threshold or contrast sensitivity function in human vision. In addition to changes in the modulation of the signals, rendering algorithms and the pixel patterns introduce sampling errors. It is possible that by assessing both the MTF and the sampling errors, one might gain significant insight into the utility of the various pixel patterns.

Traditionally, MTF-based metrics have been applied to quantify systems using cameras or a scanned sensor once the system has been constructed [16] or by quantifying the MTF of various components in a system and convolving the component MTFs to create a system MTF [13]. The influence of the human visual system can then be accounted for by computing a metric which involves both the system MTF, expressed in cycles per visual angle at the design viewing distance, and the human luminance CSF [1, 25]. While MTF metrics have been useful in the assessment of display image quality, their use typically only captures the ability of the system to transmit visible information to the user, often ignoring the artifacts that are created. However, the need to consider these artifacts in assessing overall image quality is well understood in the image quality literature [11] and recent research has demonstrated the calculation of aliasing artifacts in imaging systems, which are likely to vary for the various pixel patterns [15]. Unfortunately, the use of these methods has not yet been robustly demonstrated as an assessment method for the various pixel patterns.

8.6 Summary and Questions for Thought

Throughout this chapter, we have discussed the creation and image quality of multi-primary displays where these displays were created by adding additional colors of light-emitting elements to enhance the typical RGB triad of light-emitting elements. As we discussed, for high resolution displays (Quad HD or greater), it can be possible to leverage the fact that the additional light-emitting elements will generally produce light with a high luminance to potentially reduce the required number of light-emitting elements per display pixel. However, for lower resolution displays, it will be necessary to increase the number of light-emitting elements per pixel, which will reduce the aperture ratio of each light-emitting element in the display, reducing the efficacy of LCD light-emitting elements and the lifetime of LED light-emitting elements.

Regardless of the arrangement of light-emitting elements within the display, these multi-primary displays will require changes to the rendering algorithms which are applied to render information to the display. In certain application areas, this requirement can have significant implications. For example, computer operating systems often perform this rendering step and therefore, successful commercialization of a multi-primary display in this market might not only require the construction of an appropriate display but might require a business arrangement with operating systems companies to provide support for appropriate rendering algorithms.

Finally, we discussed the fact that the design and manufacturing of prototype displays is an expensive process and therefore it is necessary to facilitate evaluation and selection of appropriate pixel patterns and rendering algorithms through a more cost effective means other than construction of multiple display prototypes, each with a different arrangement of light-emitting elements. Three separate approaches were discussed, each of which have advantages and disadvantages while none of these methods eliminate the risk associated with adoption of a new pixel arrangement.

In this chapter, we hopefully answered a number of questions. However, we likely raised other questions which will need to be answered with time. Some questions for consideration include the following.

(1) Each of the pixel arrangements shown includes only four colors of light-emitting elements. While the literature at times illustrates the use of five or even six colors of light-emitting elements, under what conditions should we consider adding multiple colors of additional light-emitting elements?
(2) We discussed the fact that two light-emitting element per pixel arrangements could provide nearly equivalent image quality to RGB pixel arrangements for Quad HDTV displays when viewed from 3 picture heights. Is this adequate and how do we counter reviewers or critics who evaluate the displays at extremely close viewing distances?
(3) How important is the development and validation of alternative image quality assessment methods to the adoption of multi-primary displays and especially multi-primary displays with reduced numbers of light-emitting elements per pixel?

References

1. Barten PGJ (1989) The square root integral (SQRI): a new metric to describe the effect of various display parameters on perceived image quality. In: Proceedings of the SPIE 1077, human vision, visual processing and digital display, Los Angeles, CA
2. Barten PGJ (1999) Contrast sensitivity of the human eye and its effects on image quality. HV Press, Knegsel. http://doi.org/10.1117/3.353254
3. Brown Elliott CH, Higgins MF, Hwang S, Han S, Botzas A, Hsu B-S, Nishimura S et al (2008) PenTile RGBW® color processing. In: SID symposium digest of technical papers, vol 39, p 1112. http://doi.org/10.1889/1.3069331
4. Choi K, Seo S, Min B (2018). Quantitative evaluation of visual display resolution based on human visual perception. In: SID display week, pp 1–4
5. Chorin MB (2011) Performance evaluation of multi-primary color-matrix layouts for mobile displays. J Soc Inf Disp 19(2):238–245. https://doi.org/10.1889/jsid19.2.238
6. Daly SJ (1992) Visible differences predictor: an algorithm for the assessment of image quality. In: Rogoitz BE (ed) SPIE 1666, human vision, visual processing and digital display III. The International Society for Optics and Photonics

7. DePalma JJ, Lowry EM (1962) Sine-wave response of the visual system. II. Sine-wave and square-wave contrast sensitivity. J Opt Soc Am 52(3):328–335
8. Engelhardt T, Schmidt TW, Kautz J, Dachsbacher C (2014) Low-cost subpixel rendering for diverse displays. Comput Graph Forum 33(1):199–209. https://doi.org/10.1111/cgf.12267
9. Farrell J, Eldar S, Larson K, Matskewich T, Wandell B (2011) Optimizing subpixel rendering using a perceptual metric. J Soc Inf Disp 19(8):513–519. https://doi.org/10.1889/JSID19.8.513
10. Gescheider GA (1985) Psychophysics: method, theory, and applications, 2nd edn. Lawrence Erlbaum Associates Inc., Hillsdale, NJ
11. Keelan BW (2002) Handbook of image quality: characterization and prediction. Marcel Dekker Inc., New York, NY
12. Klompenhouwer MA, de Haan G (2003) Subpixel image scaling for color-matrix displays. J Soc Inf Disp 11(1):99. http://doi.org/10.1889/1.1831726
13. Kopeika NS (1998) A system engineering approach to imaging. The International Society for Optical Engineering, Bellingham, WA
14. Kranz JH, Silverstein LD (1990) Color matrix display image quality: the effects of luminance and spatial sampling. In: SID symposium digest of technical papers, pp 29–32
15. Lloyd CJ (2014) Determinants of system resolution for modern simulation training display systems. In: Image 2014 conference
16. Lloyd CJ, Basinger JD (2013) Towards an objective and affordable metric of display system resolution. In: Image 2013 conference, Scottsdale, AZ, p 9
17. Messing DS, Daly S (2002) Improved display resolution of subsampled colour images using subpixel addressing. In: International conference on image processing, New York, NY, pp 625–628
18. Messing DS, Kerofsky L, Daly S (2003) Subpixel rendering on nonstriped colour matrix displays. In: IEEE international conference on image processing, vol 2, Barcelona, Spain, pp 949–952. http://doi.org/10.1109/ICIP.2003.1246840
19. Miller ME, Arnold AD, Tutt L (2007) When is sub-sampling in RGB displays practical? In: SID symposium digest of technical papers, Long Beach, CA, pp 1146–1149
20. Miller ME, Murdoch MJ, Kane PJ, Arnold AD (2005) Image quality impact of pixel patterns and image processing algorithms for RGBW OLED displays. In: SID symposium digest of technical papers, Boston, MA, pp 398–401
21. Mullen KT (1985) Bornstein changes in brightness matches may have produced artifacts in previous isoluminant. J Physiol 359(1):381–400. https://doi.org/10.1113/jphysiol.1985.sp015591
22. Platt JC (2000) Optimal filtering for patterned displays. IEEE Signal Process Lett 7(7):179–181. https://doi.org/10.1109/97.847362
23. Primerano B, Miller ME, Murdoch MJ (2005) Method for transforming three color input signals to four or more output signals for a color display. United States Patent Number 6,885,380
24. Silverstein LD, Krantz JH, Gomer FE, Yeh Y, Monty RW (1990) Effects of spatial sampling and luminance quantization on the image quality of color matrix displays. J Opt Soc Am A 7(10):1955–1968. https://doi.org/10.1364/JOSAA.7.001955
25. Snyder HL (1985). Image quality: measures and visual performance. In LE Tannas Jr (ed) Flat-panel displays and CRTs. Van Nostrand Reinhold, New York, NY, pp 70–90
26. Spindler JP, Hatwar TK, Miller ME, Arnold AD, Murdoch MJ, Kane PJ, Van Slyke SA et al (2006) System considerations for RGBW OLED displays. J Soc Inf Disp 14(1):37. https://doi.org/10.1889/1.2166833
27. Torgeson WS (1958) Theory and methods of scaling. Wiley, New York, NY
28. Zhang X, Silverstein DA, Farrell JE, Wandell BA (1997) Color image quality metric S-CIELAB and its application on halftone texture visibility. In: Proceedings IEEE COMPCON 97. Digest of papers, pp 44–48. http://doi.org/10.1109/CMPCON.1997.584669
29. Zhang X, Wandell BA (1997) A spatial extension of CIELAB for digital color-image reproduction. J Soc Inf Disp 5(1):61. https://doi.org/10.1889/1.1985127
30. Zheng X (2014) Optimization of sampling structure conversion methods for color mosaic displays, University of Ottawa

Chapter 9
The Multi-primary Advantage

9.1 An Overview

In the early chapters of this book, we looked at various aspects of the display environment and current display technologies. In the last few chapters we discussed processing and design considerations for multi-primary displays. Throughout the last couple chapters, it has likely become clear that the additional primaries add complexity to the selection of appropriate pixel patterns, display design, and image processing beyond those present in a typical RGB display. So why should we consider the use of multi-primary displays given this added complexity and the corresponding cost?

We have discussed a few advantages of multi-primary displays in Chaps. 7 and 8. We discussed the fact that adding primaries outside the RGB gamut can expand the color gamut of the display. Gamut expansion of course has the potential to improve image quality. We discussed the fact that gamut expansion from the addition of primaries cannot be completely understood in a two dimensional plot as it can also increase the ability to present higher luminance colors. Finally, we discussed the fact that for very high resolution displays (e.g., Quad HD and higher) we might take advantage of the characteristics of the human visual system to construct displays that only require two light-emitting elements per pixel, rather than the three RGB light-emitting elements required in a traditional display.

With this knowledge as a backdrop, we will discuss a few very specific potential advantages of multi-primary displays in this chapter. These include power reduction and the potential reduction of the impact of viewer metamerism. Each of these advantages are important in today's displays, which must provide a large color gamut to compete successfully in a display space that is already occupied by very high quality displays. We will explore the possibility of using multi-primary displays to modify the signal to the ipRGCs. In the final section of this chapter, we will recognize that multi-primary displays can additionally be necessary to enable certain classes of displays. We have already discussed the fact that the use of RGBW has enabled the commercial production of large format OLED displays. In this example, we will look specifically at field-sequential color displays and the reduction of color breakup in these displays through the use of additional color fields.

9.2 Power and Lifetime Effects

One of the often-touted advantages of multi-primary displays is a reduction in power consumption which occurs as this technology is adopted. This section will illustrate this reduction in power. It will also recognize a number of factors which influence the amount of power that is saved. This power savings can vary by display format and the primaries that are adopted and their placement. In this section, we will discuss a few different display configurations, including addition of white, yellow, and cyan additional primaries to form multi-primary displays.

9.2.1 RGBW for Filtered White Emitters

As an initial illustration, we will begin by exploring the potential power savings provided by RGBW for a filtered white emitter. Among the factors which affect the power consumption of this class of displays are the bandwidth of a white emitter and the bandwidth of filters used to filter this white emitter. This section will explore each of these in turn to illustrate the conditions under which adoption of a multi-primary display including a filtered W light-emitting element will result in significant power savings.

> Power savings in color-filter-based RGBW displays varies based upon the spectrum of the white light source and the color filters which are selected.

For this illustration, we will apply each of the spectra shown in Fig. 9.1. Note that this figure shows two spectra. The first is a broadband white, indicated by the solid line, which is taken from the OLED literature [6]. This white spectra was produced by stacking several OLED light-emitting layers, which generally represent blue, green, yellow, and red emission in single OLED device. As we apply color filters to this emission spectra, we can expect to see a significant portion of the light that is produced to be filtered from the spectral areas between the RGB emitters as well as reduction of peak emission in the areas where the color filters transmit significant energy.

The second spectra, indicated by the dashed line in Fig. 9.1, is a theoretical narrow bandwidth white emitter. It is formed by three emitters each having a Gaussian light distribution with a full width at half maximum intensity of 30 nm. These emitters

9.2 Power and Lifetime Effects

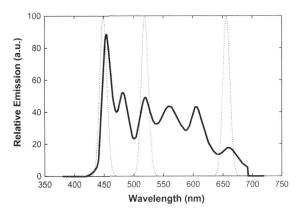

Fig. 9.1 Example broadband white emitter spectra, including a broadband white emitter designed to optimize white efficacy (solid bold line) and a narrowband white emitter designed to maximize energy transmission through the selected color filters (dotted line)

might represent idealized LEDs or a backlight formed from blue LEDs which stimulate emission from red and green quantum dots. In this case, the peak emission of these emitters have been carefully selected to be near the peak transmittance of the color filters that we will apply in this example.

It might be noted that these particular emission spectra were selected as one might assume that color filters will have the largest impact on the transmission of a broadband white emitter, but would have significantly less impact on the transmission of light from a well-designed white emitter with very narrow bandwidth emitters. Therefore, it is a reasonable assumption that these conditions generally represent a white emitter (broadband) in which the application of RGBW will have a very significant impact and a white emitter (narrowband) in which the application of RGBW will have the least possible impact.

One should note that the total energy expressed in the spectra shown in Fig. 9.1 are not equal. The area under the narrowband spectra is obviously smaller than the area under the broadband spectra. However, for the analysis we will perform, this fact is not relevant as the analysis will utilize the proportion of the energy transmitted to the human eye after color filters are applied to the total available energy present in the underlying white spectra. It is also worth noting that the efficacy of these white emitters are not similar either. The broadband spectra includes a significant amount of energy around 560 nm, which is the peak sensitivity of the human eye. Although such a peak could have been included in the narrowband spectrum to improve the efficacy of this white, such a peak was not included as the color filters would have blocked much of this energy in the RGB display equivalent. Therefore, the current analysis really attempts to compare a broadband spectrum which was optimized to improve the efficacy of the white to the human eye to a narrowband spectrum which was designed to maximize transmittance through the color filters. This is a design solution that one must make when designing the underlying white spectrum. However, as this particular analysis is focused on energy transmission through the color filters, the current conditions were selected to clearly demonstrate among the

Fig. 9.2 Comparison of typical LCD color filters (solid lines) and narrowband OLED color filters (dashed lines)

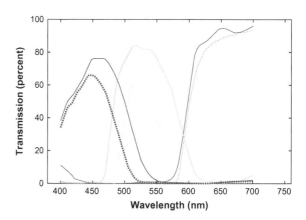

largest effects of the white emitter spectrum on the energy transmission through color filters, regardless of the efficacy of the white emitter.

We can filter these white emitters with any number of color filters. In this example, we will leverage published color filters. Specifically, we will apply a set of color filters originally designed for use in a typical LCD and a set of color filters designed to provide a larger color gamut for a broadband white OLED emitter. These color filters are depicted in Fig. 9.2. Each of these color filter spectra have been discussed by Hamer and colleagues [6].

As we have discussed, two changes often occur as the bandwidth of especially blue and green color filters is reduced. Naturally, the bandwidth is reduced. This permits purer, more saturated colors to be formed. However, a side effect of the process used to narrow the emission bands is a reduction in the peak transmission of the color filters. Each of these changes are illustrated in the emission spectra shown. When applying RGB displays, narrowing the bandwidth of the color filter will substantially increase the power consumption of the final display.

> RGB displays employing a white backlight and color filters typically trade color saturation for significant increases in power consumption.

Table 9.1 illustrates the effect of the bandwidths of the two emitters and the changes in color filters on the percent of the white light energy that is transmitted for each emitter and each set of color filters. As expected, less of the energy from each emitter is transmitted by the narrow bandwidth OLED color filters than the broader bandwidth LCD color filters. Generally, less of the energy from the broad bandwidth white emitter is transmitted by either color filter set than is transmitted by the narrow bandwidth emitters. This is not true in all cases, however, as the proportion of the broad bandwidth white emitter transmitted by the broad bandwidth green LCD filter is actually greater than proportion of energy transmitted by the narrow bandwidth

9.2 Power and Lifetime Effects

Table 9.1 Percent of white light emission transmitted by the selected color filters. Note that these percentages are shown for both broad and narrow bandwidth emitters depicted in Fig. 9.2 and for the broader bandwidth LCD color filters, as well as the OLED Color Filters depicted in Fig. 9.3

Emitter	Filter	Red	Green	Blue
Broad bandwidth	LCD	29.6	36.6	20.8
	OLED	19.5	16.2	18.4
Narrow bandwidth	LCD	32.7	28.0	31.7
	OLED	22.9	20.0	29.8

emitter. This increase in transmission likely also signals a significant reduction in color purity.

Using the white emitters shown in Fig. 9.1, the color filters shown in Fig. 9.2 and the image distribution discussed at the end of Chap. 3, it is possible to compute the relative power consumption of RGB and RGBW displays where these displays have either white emitter shown in Fig. 9.1, coupled with either no color filter, the narrow bandwidth color filters or the broad bandwidth color filters. This is accomplished by computing the relative power consumption of each primary, computing the power consumed by each possible code value in the image set and then weighting this relative power consumption by the frequency of occurrence for that individual code value based on the image distribution. For conditions where a white light-emitting element is included, it is assumed that all of the energy from the white emitter passes through the W emitter and that all of the neutral portions of the color within the image set are produced by the W emitter. That is the basic algorithm discussed in Sect. 7.2.2.

As shown in Table 9.2, if no color filters are used, the power consumption is normalized to a value of 1. Because the LCD color filters each block slightly more than 1/3rd of the available light from the emitter, the RGB version of the display has a power consumption that is a little greater than 3, values of 4.53 for the broad bandwidth emitter and 3.57 for the narrow bandwidth emitter, to be precise. As the color filters are narrowed, they block more of the light. Therefore, the OLED color filters, while providing a significantly larger color gamut, also increase the power consumption to a value that is 6.72 and 5.29 times the power consumption of a display producing the same luminance without any color filters for the broad and narrow bandwidth emitters, respectively. Therefore, adding the color filters, and especially the narrowband color filters, significantly increase the power consumption of the display. Of course this is not unexpected. Color filters permit the display to produce color and this color comes at a cost of power consumption in a display employing color filters. Further, the narrower the bandwidth of the color filters, the more saturated the final light emission becomes and the larger the power consumption penalty is that is incurred.

Table 9.2 Relative power consumption as a function of color filters and display format. LCD refers to broadband color filters and OLED refers to narrowband color filters as depicted in Fig. 9.3

Emitter	Format	No filters	LCD	OLED
Broad bandwidth	RGB	1.00	4.53	6.72
	RGBW	1.00	1.59	1.99
Narrow bandwidth	RGB	1.00	3.57	5.29
	RGBW	1.00	1.49	1.69

> The power savings provided by RGBW is significantly greater for displays which employ narrower bandwidth, less efficient color filters than for displays which employ broader bandwidth, more efficient color filters.

However, as shown in Table 9.2, assuming that the color of the white emitter is at the white point of the display, it is possible to also estimate the relative power consumption of a RGBW format display. In this example, we assume that all neutral energy is removed from the RGB light-emitting elements and produced by the W light-emitting element. As shown, for the LCD color filtered display, the relative power for the display ranges from 1.49 for the narrow bandwidth emitter to 1.59 for the broad bandwidth emitter. As such, the power consumption of the RGBW format is 35–40% of the power consumption of the RGB format. The power savings provided by RGBW becomes even greater as narrower band color filters are adopted. For the narrow-bandwidth OLED color filters, the RGBW format has a relative power consumption of 1.69–1.99 times the unfiltered display. This value is approximately 30% (29.6–32.0%) of the power consumption of the RGB display. This table illustrates the fact that as the bandwidth of the color filters are narrowed, the power consumption benefit of RGBW over RGB increases. This benefit exists regardless of whether the emitter is narrow or broad bandwidth in nature.

> Power savings available from the use of RGBW in a color-filtered display is significantly different for emissive than transmissive or reflective displays. This difference is further dependent upon the pixel pattern and the relative size of the apertures for transmissive or reflective displays.

Thus far, we have assumed that we are able to harvest all of the additional light which passes through the additional W emitter without further reducing the energy which passes through the RGB light-emitting elements. This assumption is generally correct within emissive displays. However, this assumption is not true for LCDs or other transmissive or reflective displays where reducing the size of the light-emitting elements reduces the size of the aperture through which light can pass or

9.2 Power and Lifetime Effects

be reflected and, thus, reduces the efficiency of the RGB light-emitting elements. As a rough approximation to illustrate this effect, let's assume that regardless of the number of light-emitting elements per pixel, that 80% of the area of the display panel is available for light transmission. In reality, this value scales with the number of elements, decreasing as the number of elements is increased. However, this effect is less than the effect of apportioning the available light-emitting area to the light-emitting elements. With this assumption, the available area for light modulation per light emitting element varies with the number of light emitting elements with values of 40% for two elements per pixel, 26.7% for 3 elements per pixel and 20% for three elements per pixel. In this example, adding the W emitter, requiring 4 elements instead of 3, offsets any increase in efficiency somewhat as this change reduces the efficiency to about 75% of the gain assumed earlier. Of course, if an arrangement of light-emitting elements can be adopted which reduces the number of light-emitting elements to 2 instead of 3, an additional 50% improvement in power consumption can be harvested in these displays.

Thus far, we have only discussed power consumption, not display lifetime. Lifetime of emissive displays can be altered significantly by the inclusion of additional elements. While the aperture ratio and the power consumption of LCDs is modified as emitters are added or subtracted, similar changes in the aperture ratio of the emitters in an emissive display results in changes in current density within emissive pixels. Decreasing the aperture ratio increases the current density within each light emitter for a constant current and the lifetime of that emitter is decreased. Therefore, adding elements can reduce the lifetime of the display if a field of constant color (e.g., a full field of red) is presented as compared to the lifetime of an RGB display. This difference occurs as the size of the light-emitting elements will be smaller in RGBW displays than in an RGB displays having the same pixel size. Therefore, the current density (i.e., current per area) will be increased when the current is the same through the red OLEDs in each display format. On average, however, this lifetime effect is offset because the average current is typically reduced by the inclusion of the W light-emitting element which reduces the average current by a factor that is large enough to overcome the loss in emission area. In fact, it has been shown that including a W emitter in an emissive RGBW display can, on average, increase the average lifetime of the RGBW display as compared to a similar RGB display [20]. However, this relationship is complex and depends upon the content which is actually displayed.

The relationship between lifetime of an emissive display and the addition of light-emitting elements to form a multi-primary display is complex and highly dependent upon content.

9.2.2 RGBW for Unfiltered Emitters

Earlier, we discussed measurements from an iLED display which illustrated that a single white iLED could produce a greater luminance output than a combination of a red, green, and blue iLED driven at the same current. This fact is perhaps counter intuitive, as the same number of electrons are being inserted to each of the four iLEDs and yet the white iLED is producing as much or more useful light as the sum of the red, green, and blue iLEDs. This phenomena, however, occurs because the white iLED is composed of a blue iLED coated with a yellow phosphor. Because of the presence of the yellow phosphor, the white iLED produces a significant number of photons which are near the peak sensitivity of the human eye. Therefore, the efficacy of the iLED is significantly greater than that of the RGB iLEDs due to the improved response of the eye to the light that is generated. Further, blue iLEDs are extremely efficient at converting electrons to photons and the yellow phosphor is extremely efficient at converting blue photons to longer wavelength photons. Therefore, these iLEDs can produce a greater number of photons than longer wavelength iLEDs for the same power consumption. Similar relationships can exist in OLED and possible other emitting technologies [5, 21].

With this phenomena, one can then ask whether inclusion of a white light-emitting element can significantly improve the power consumption of unfiltered emissive displays. To evaluate this question, one can simulate a display having a white luminance efficacy which is some multiple of the white efficacy of a triplet of RGB primaries. Once again, this analysis will include the image distribution which was discussed in Chap. 3 to help us understand the proportion of neutral energy which can be converted from the RGB emitters to the W emitter. The results as the efficacy of the W emitter is varied from 0.25 to 3 times the efficacy of white produced by the RGB emitters is shown in Fig. 9.3. The analysis performed to generate this figure is once again, a relative analysis. The analysis assumes that whatever the efficiency of the RGB emitters, the color white is produced by the RGB emitters and consumes 1 power unit. We might consider if the RGB elements each consume the same amount of power then they each consume 0.33 power units. While this is not necessary, it is useful to understand that the average power consumption of the R, G, and B light-emitting elements is 0.33 within this assumed device.

> Power consumption of displays having patterned RGBW emitters can be significantly lower than the power consumption of displays having patterned RGB emitters as long as the efficacy of the W emitter is significantly higher than the average efficacy of the RGB emitters.

The addition of the W emitter changes the power consumption of the display relative to this value. If the W emitter creates an equivalent white to the white produced by the RGB elements, the W element must consume three times the average power consumption of the RGB elements to have the same efficacy. In this assumed

9.2 Power and Lifetime Effects

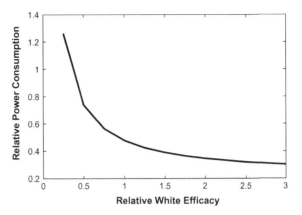

Fig. 9.3 Power consumption of a RGBW display relative to a comparable RGB display as a function of the relative W power efficacy, where the W power efficacy is calculated as the ratio of the W efficacy to the efficacy of white created in the comparable RGB display

display configuration, if a W emitter is added which permits the average image to be produced with 0.5 power units, this display has twice the efficacy as the comparable RGB display. Examining Fig. 9.3, the lowest white efficacy which is assumed for the W emitter is 0.25. Under this condition, the efficacy of the W emitter is lower than the average efficacy of the individual RGB emitters. Therefore, as Fig. 9.3 illustrates, the relative power consumption of the display, shown on the y axis, is greater than 1 for this condition. However, power savings is achieved as the efficacy of the W emitter increases to values higher than 0.33. As shown, including a W emitter with an efficacy as low as 0.5 reduces the power consumption of the display to 80% of the power consumed by the comparable RGB display. Including W emitters with an efficacy of 1 can reduce the power consumption of that produced by the RGB display. Further, adopting a W emitter with an efficacy of 1.25, approximately the value discussed from earlier measurements of R, G, B and W iLEDs, reduces the power consumption of the display to about 42% of the power consumption required for the comparable RGB display. Therefore, the inclusion of a W emitter can significantly improve the power consumption of unfiltered emissive displays.

The addition of unfiltered (e.g., white) elements to a display having a white backlight can improve the overall efficiency (i.e., ratio of radiant power to input power), and thus the efficacy (ratio of luminance output to input power), of the display. However, the reduction in power consumption from multi-primary displays is not restricted to filtered displays. The power consumption of unfiltered emissive displays can also benefit significantly from the addition of white or other colors of light-emitting elements which have a higher efficacy than the efficacy of R, G, or B emitting elements.

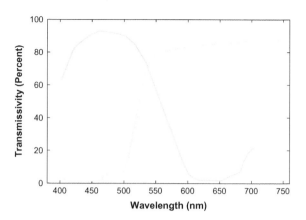

Fig. 9.4 Color filter transmissivity for a typical yellow color filter, indicated by the solid yellow line. The filter transmissivity for a typical cyan color filter is also shown, indicated by the solid cyan line. Color filters adapted from [12]

9.2.3 RGBY for Filtered Emitters

As we discuss power savings, yellow light-emitting elements are the next most important primary to discuss. These light-emitting elements, similarly to the W light-emitting elements, often benefit from both high efficiency color filters and improved sensitivity of the human eye.

Yellow color filters, much like red color filters, must only reduce the transmission of short wavelength light. Therefore, it is possible to design yellow color filters that both have a broad bandwidth and transmit relevant wavelengths of light at efficiencies greater than 80%. Each of these attributes are illustrated by the yellow color filter spectra shown in Fig. 9.4. As shown, the transmittance of this color filter is greater than 80% for wavelengths of light within the visible spectrum longer than about 550 nm. Note that this range can include green, yellow, and red light emission. Therefore, when a yellow filter is placed over a white emitter, it can transmit a large portion of the available energy. As we have now mentioned several times, the eye's sensitivity is greater for light having a wavelength near 560 nm than for any other light. This light is perceived by the human as a greenish-yellow light. Therefore, it is possible to develop yellow light-emitting elements having a higher efficacy than any other light-emitting elements.

Employing these yellow light-emitting elements within a display, we will generally replace light which is generated by the green and red light-emitting elements during color rendering. The filtering efficiency for the red light-emitting element is generally good and the efficacy of the green light-emitting elements are also generally good, therefore, it is not extremely clear that the addition of a yellow light-emitting element will significantly reduce the power consumption of a display. In some technologies, yellow light emission has other significant advantages. For example, in OLED devices high efficiency yellow-light emitting triplet materials are known which have exceptional efficiency [21].

Table 9.3 Relative efficiencies and efficacies for each primary in the RGBY display. All efficiency values are relative to the efficacy of the unfiltered white

White emitter	Efficiencies				Efficacies			
	Red	Green	Blue	Yellow	Red	Green	Blue	Yellow
Three peek	29.8	20.0	22.9	42.1	9.0	50.9	5.3	43.6
Four peek	22.3	23.7	17.2	51.2	4.1	42.1	2.5	63.0

In displays employing color filters, elements composed of broadband secondary color filters can provide much more light transmission than narrowband primary color filters.

To illustrate the relative power consumption impact of the addition of a yellow emitter, we will employ a color filter solution. Specifically, we will assume a white emitter spectra indicated by the narrowband spectra shown in Fig. 9.1. For further illustration, we will include a yellow emitter in this spectrum having the same bandwidth as the peaks shown but having a peak at 555 nm, which corresponds to the peak sensitivity of the human eye. Further, we will assume the use of the OLED RGB filters from Fig. 9.2. An additional yellow, light-emitting element will be assumed having the same underlying emission spectra, but employing the yellow color filter shown in Fig. 9.4. The relative efficiencies and efficacies of each resulting emitter is shown in Table 9.3. Note that when calculating the relative efficacy, we are including both the relative efficiency of the energy-emission and the sensitivity of the human eye. The efficacy is calculated as the proportion of the luminance of the final emitter to the luminance of the underlying white spectra, assuming that the efficiency is the same for each white spectra.

As Table 9.3 shows, the efficiency of the yellow emitter is much higher than the efficiency of the red, green, or blue emitters due to the broad bandwidth of the yellow color filter. Further, when the yellow emission peak is added to the white, the efficacy of this primary is much greater than the efficacy of the other emitter's due to the increased sensitivity of the human eye to illumination in the area around 555 nm.

It is useful to understand the use of the primaries for forming the white point of the display. Table 9.4 shows the relative luminance required to form a D65 white from the appropriate color combinations. These include the RGB combination on the left and the RGBY combination on the right, as the yellow primary is actually used to replace the green primary when forming white in this particular example. In each of these cases, one can see that the red luminance is reduced significantly while the green luminance is eliminated. The vast majority of this luminance is replaced with luminance from the yellow primary, which as we noted before also has a higher efficacy, particularly in the 4 peak configuration.

Table 9.4 Aim relative luminance values when forming a 100 cd/m², D65 white for both the RGB case and the RGBY case where the luminance of the Y emitter replaces all of the energy from the G emitter

White emitter	RGB				RGBY			
	Red	Green	Blue	Yellow	Red	Green	Blue	Yellow
Three peek	26.2	70.2	3.6	0	8.0	0	3.5	88.5
Four peek	20.8	75.3	3.9	0	11.0	0	4.0	85.0

Table 9.5 Relative power consumption as a function of emitter and display format

Emitter	Display format	
	RGB	RGBY
Three peek	5.0	3.6
Four peak	3.8	2.7

Assuming that the luminance of the yellow should be set to the luminance of a point on the red/green gamut boundary, corresponding to the point where a line drawn from the display white point to the yellow primary intersects this gamut boundary, one can establish the aim relative luminance of the yellow primary. This primary can then be used with the algorithms discussed in Sect. 7.2.3 to estimate the power consumption of the RGBY display relative to the RGB display. The resulting relative power consumption of the resulting display is shown in Table 9.5.

As shown in Table 9.5, the power consumption of the RGBY display is significantly lower than the RGB display. The ratio of the RGBY power consumption to the RGB power consumption is approximately equal with the RGBY format requiring around 71–72% of the power consumption of the equivalent RGB display. It should be noted that by adding the 4th peak to the emitter, the relative efficacy of the display improves, however, this solution results in a loss of color saturation, as illustrated in the next section.

9.2.4 RGBYC for Filtered Emitters

In color filter systems, cyan color filters also tend to be broadband in nature, permitting a relatively large proportion of the light emitted by an emitter to pass to the human viewer. Therefore, pixels employing a color filter over an emitter with multiple emission peaks, particularly in the blue and green regions of the visible spectrum, will be much more energy efficient than a single blue or green emitter. Further, the eye is more sensitive to cyan light than to blue. Therefore, the addition of a cyan

9.2 Power and Lifetime Effects 195

Table 9.6 Relative efficiencies and efficacies for each primary in the RGBC display. All efficiency values are relative to an unfiltered white, efficacies are relative to the unfiltered 3-peak white

White emitter	Efficiencies				Efficacies			
	Red	Green	Blue	Cyan	Red	Green	Blue	Cyan
Three peek	29.8	20.0	22.9	58.7	2.3	12.7	1.3	18.7
Four peek	22.3	23.7	17.2	57.2	1.7	17.7	1.0	26.3

emitter, particularly in combination with a yellow emitter, can provide significant power advantages.

In addition to the yellow color filter, Fig. 9.4 also depicts a typical cyan filter. As shown, this cyan color filter also has a high transmission, providing a transmissivity of 80% or greater for wavelengths between approximately 400 and 520 nm. This high transmission over a broad bandwidth enables the relatively high efficiency of cyan filtered pixels. The inclusion of light in the green portion of the visible spectrum improves the efficacy of the resulting pixel. Table 9.6 shows the relative efficiencies and efficacies of each primary in a display similar to the one shown in Table 9.3 where the only difference is the replacement of the yellow color filtered pixel with the cyan color filtered pixel.

As shown in Table 9.6, the cyan pixel has an efficiency which is approximately twice the efficiency of the red, green, or blue pixels. The efficacy of the cyan pixel is also much greater than the efficacy for the red or blue elements and slightly greater than it is for the green element. Comparing the results in Table 9.6 to those in Table 9.3, as the efficiency is higher for the cyan element relative to the RGB elements, the difference in the efficacy of this particular cyan pixel is even higher than the difference in efficacy between the particular yellow element and the RGB elements in the previous example. Therefore, if only one additional light-emitting element is to be added to a display to improve power efficiency, this example would appear to indicate that the addition of the cyan element might be more desirable than the addition of the yellow emitter. This decision is actually much more complex than indicated. Other factors, including the white point of the display and the location of the other primaries within the color space each contribute to this decision. Generally, it is more desirable to add a yellow than a cyan element. However, adding both a yellow and cyan element can significantly reduce the power consumption of a display.

Table 9.7 shows the luminance ratios of the red, green and blue primaries when D65 white is formed from the RGB primaries in this example as well as the red, green, blue, yellow and cyan luminance ratios when the same D65 white is formed from the RGBYC primaries, where as much of the energy is possible is formed from the yellow and cyan primaries. As shown, the blue and green primaries are not used to form white in this example. Fifteen to 20% of the energy is provided by the red, while the remainder of the light is formed from the yellow and cyan primaries. Even

Table 9.7 Aim relative luminance values (in cd/m^2) when forming a 100 cd/m^2, D65 white for both the RGB case and the RGBYC case

White emitter	RGB			RGBYC				
	Red	Green	Blue	Red	Green	Blue	Yellow	Cyan
Three peek	26.2	70.2	3.6	16.4	0.0	0.0	47.6	36.0
Four peek	20.8	75.3	3.9	19.3	0.0	0.0	8.4	72.3

Table 9.8 Relative power consumption as a function of emitter and display format

Emitter	Display format	
	RGB	RGBYC
Three peek	5.0	3.4
Four peak	3.8	2.6

the amount of light demanded by the red primary is reduced in the RGBYC condition as compared to the RGB condition.

> The changes in display power consumption which are provided by adding a power efficient primary is highly dependent upon the other light-emitting elements being considered for the display.

As we will not address this anywhere else, it is also worth noting that one could additionally add a magenta pixel to a display. If this element includes color filters, magenta color filters must have two transmissive peaks, improving the efficiency of this color filter over the blue and red color filters. However, each of these peaks will often be narrow and therefore the power advantage gained is not as great as it is for yellow and cyan filters as the peak transmittance will often be reduced, particularly for the blue portion of the light. Further, the efficacy of the magenta light is somewhere between that of the blue and red light. Therefore, the addition of this light-emitting element will provide a relatively modest reduction in power consumption and this tradeoff must be considered with respect to the increased number of light-emitting elements which it will require.

Returning to our discussion of RGBYC color filtered displays, the algorithms discussed in Sect. 7.2.3 can be applied to estimate the power consumption of the RGBYC display relative to the RGB display. The resulting relative power consumption of the display is shown in Table 9.8.

As shown in Table 9.8, the results for the RGBYC display are nearly equivalent to that for the RGBY display. This occurs in this example despite the fact that the neutral color is formed from the yellow, red, and cyan emitters. Had the cyan or

9.2 Power and Lifetime Effects

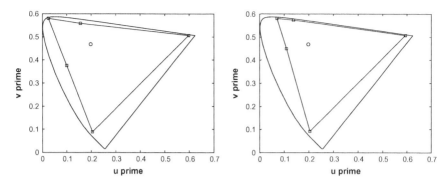

Fig. 9.5 Color gamut for the RGBCY displays considered in this example. Three peak white emitter is shown on left and four peak white emitter shown on right

yellow been shifted towards the blue or red, such that the display neutral would have been formed by the cyan, green, and yellow emitters, the addition of the cyan emitter would have likely had a larger effect even though the efficacy of the individual emitters would have been reduced. This example illustrates the importance of the selection of primaries in light of the content that is to be produced.

Additionally, Fig. 9.5 shows the loss of color gamut which occurs as one adopts the 4 peak rather than the 3 peak emitter. The green is very saturated when applying the 3 peak emitter but becomes much less saturated as the 4th peak is added. Of course the degree of this change will depend upon the exact placement of this peak within the existing spectra and the tradeoff between power consumption and color gamut is left to the discretion of the display designer. It is also interesting that because the cyan and yellow filters used in this example were quite broad, the uniform chromaticity coordinates for these primaries essentially lie on the RGB boundary. That is, the use of these emitters do not improve the color gamut of the display system to any measurable degree. Once again, this illustrates another tradeoff in these displays, providing the potential for increases in color gamut and power efficacy to be exchanged for one another.

9.3 Observer Metamerism

Throughout this book, we have used the CIE 1931 color matching functions in the calculation of each of our color metrics and to support the image processing algorithms that have been proposed. These color matching functions provide a method for reducing complex spectra into a luminance and two chromaticity values to represent each color. Underlying this work are two potential assumptions. The first assumption is if we have a drive signal that is linear with luminance, which is calculated using the $V(\lambda)$ function or the equivalent from the color matching functions, the perceived relationship between input stimulus and output light intensity is linear and color additivity holds. That is, Grassman's laws, which we mentioned in Chap. 2, holds and two objects, even though they have different spectral reflection or emission, can appear the same. This creates what is referred to as a metameric match and, thus, Metamerism is defined as the ability for a user to perceive the color of two objects having different spectral composition as the same. Based on this assumption, we can utilize the color matching functions to determine the relative differences in perceived color on a display having a single set of primaries. The second assumption that was implied throughout this book is that we could apply the chromaticity values to understand the degree of match not only between two colors on a display with a set of primaries but we could also apply the luminance and chromaticity values to understand the degree of match between colors on a display with one set of primaries and either colors on a different display having a second set of primaries or a real world scene containing objects colored by embedded dyes or pigments. The implication of this second assumption is that any two spectra which result in the same luminance and chromaticity values would appear identical to all people, regardless of the underlying spectral shape.

Each of these assumptions have been important as they simplify our analysis and provide an approximation to reality. However, the later assumption is not accurate. The color matching functions represent the visual performance for a "standard" observer but do not represent the visual performance for every, or perhaps, any observer. Instead, people's visual systems are somewhat different and therefore we do not all perceive light and color the same. Instead the color of lens in our eyes, the color of the pigment at the back of our eyes, the photo pigmentation of our individual cones, and the projection of the visual pathways into our brains all differ between individuals [1]. As a result, different individuals can perceive two colors which have identical luminance and chromaticity values but different underlying spectral composition as having different color and luminance. This mismatch in color for colors having equivalent luminance and chromaticity values is referred to as metameric failure. Further, the degree of the difference which is likely to be perceived depends upon the shape of the underlying spectra.

There are two issues with non-metameric matches which are important in multi-primary displays. First, it is possible through selection of the primaries and the replacement strategies that two individuals will perceive two colors which were intended to be matched but rendered with different primary colors to be perceived differently than desired [3]. The second problem, and the one we are most concerned with in this section, is the fact that an image rendered on a display might have identical luminance and chromaticity to a reference image in the real world or on another display and yet be perceived as having different colors [2].

It turns out that this problem is very relevant to multi-primary displays. Generally, the display industry is driving towards the use of more highly saturated color primaries. On the surface, this effort makes perfect sense. Increasing the saturation of the color primaries increases the color gamut of the display and increases the ratio of the number of colors a display can produce to the number of colors a human can perceive. In side by side comparisons, users value this increase in color saturation,

9.3 Observer Metamerism

showing strong preference for displays with higher color saturation as long as the color rendering path maintains the hue of the original image and does not oversaturate certain memory colors, such as the color of human flesh.

This drive towards a larger display color gamut through the use of narrower and narrower band primaries has negative consequences. As discussed in the previous section of this chapter, in RGB displays the use of narrow bandwidth primaries, particularly when using color filters, can dramatically decrease the efficiency of the display. However, this drive towards narrower bandwidth primaries also increases the likelihood and magnitude of metameric failure [16, 18]. To understand this effect, we only need to consider that the sensors in our eye are differentially sensitive to the spectrum of the light to which the eye is exposed. Each sensor in the human eye has a bandwidth range where the sensitivity decreases rapidly as a function of wavelength. If the sensors are exposed to a light with a very narrow bandwidth which emits light in the wavelength range over which sensitivity decreases rapidly, a slight change in the wavelength of the filter will dramatically change the stimulation of the eye. As such, the perceived difference in sensation will be large in response to this slight change in wavelength. However, if the light has a very broad bandwidth or multiple peaks, it is unlikely that all of the light entering the eye will be influenced by these minor shifts in the eye's response and the perceived change due to such a shift will be very small.

> The addition of broad bandwidth primaries can significantly reduce the likelihood and magnitude of metameric failure between displays or a display and a real world color.

It is, therefore, not only important that primaries having narrower bandwidth will be more likely to result in larger metameric failures, but, stated from an inverse point of view, metameric failure can be minimized through the use of broad bandwidth or multiple-peaked primaries, including the broad bandwidth white, yellow, or cyan primaries discussed in the previous section. In fact, under the right conditions, color metamerism failure can be reduced through the use of an increased number of narrow bandwidth primaries [10]. The use of multi-primary displays then permits us to form most colors using these broad bandwidth primaries, which minimizes the likelihood and magnitude of metameric failure. However, by including very narrow bandwidth RGB or other narrow bandwidth primaries in the display we retain the ability to produce very saturated colors. While the use of the pure primary colors can still result in metameric failure for very saturated colors, the use of the broad bandwidth primaries in the formation of less saturated colors reduces the likelihood and magnitude of metameric failure for the vast majority of colors in the display. Although we could demonstrate this fact here for a given set of primaries, this fact has been clearly discussed by Long and Fairchild [10] and the reader is referred to this paper for further details.

It is also useful to understand that metameric failure of the type discussed here actually occurs for two separate reasons. One of them is the difference between individuals, as noted earlier. The other reason is that the CIE 1931 standard observer does not accurately represent the average observer. Therefore, image processing approaches have been developed to attempt to reduce the magnitude of metameric failure for the average observer [2]. However, these approaches do not overcome the issues associated with individual differences. As such, multi-primary displays provide a significant advantage when creating displays with highly saturated primaries.

9.4 Modifying Stimulation of the ipRGCs

In Chap. 2, we discussed the presence of the intrinsically photosensitive retinal ganglion cells (ipRGCs) in the human retina. These cells are believed to provide a signal which aids photo-entrainment and regulation of the circadian rhythms. In some displays provided today, the energy provided to these cells is modified by adjusting the color temperature of the display throughout the day, providing an increased color temperature around midday and a reduced color temperature in the evening, somewhat tracking the color temperature of the sun. In this approach, the stimulation of these sensors is increased early in the day but reduced in the evening, helping to encourage melatonin production and, ideally, increasing drowsiness near bed time. Unfortunately, some users find the yellowing of the light associated with lower color temperatures unpleasant.

In this section, we will discuss the use of a multi-primary display to alter the energy provided to the ipRGCs without modifying the color temperature. As such, we can construct a multi-primary display which might be useful in photo-entrainment and circadian rhythm regulation. However, the science in this area is still developing, and at the current time the true effects of illumination and the role that the rods and cones play in photo-entrainment and circadian rhythm regulation is uncertain. For the purposes of this discussion, we will assume that these biological phenomena are affected solely by the ipRGCs and the rods and cones are not involved in regulation of these biological phenomena. The reader should understand, however, that this is likely a gross simplification of the true process and evidence exists which suggests that at least the Beta and Rho cones play a significant role in regulation of these biological phenomena [19]. Further, we will assume that the sensitivity function, as published by Lucas and colleagues [11] effectively characterizes the light sensitivity of the ipRGCs.

In this display, we will assume that 4 emitters exist. Three of these we will refer to as red, green, and blue light-emitting elements. In this display, we will purposefully separate the blue and green light emitting elements from the peak of the ipRGC sensitivity curve. We will then assume the presence of a fourth light-emitting element which we will refer to as cyan. This fourth emitter will have a peak output very near the peak sensitivity of the ipRGCs. Spectra for four such light-emitting elements, as well as the ipRGC sensitivity curve, is shown in Fig. 9.6.

9.4 Modifying Stimulation of the ipRGCs

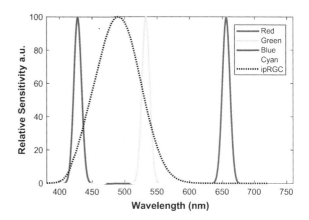

Fig. 9.6 Illustration of the normalized energy output of the selected red, green, blue and cyan light-emitting elements, as well as the sensitivity of the ipRGCs

As shown in Fig. 9.6, if one integrates the energy under the area of the ipRGC sensitivity curve for each of the emitters, the energy will be the greatest for the cyan light-emitting element and will be reduced for each of the remaining light-emitting elements. We can then consider forming the white point of a display using one combination of three of four of the light-emitting elements from this combination. Specifically, in one configuration, we will rely primarily upon the red, green, and blue light-emitting elements from the display. In a second configuration, we will rely less on light from the blue and green light-emitting elements and rely much more heavily on the cyan light-emitting element to form the white point of the display. The two relevant color triangles, as well as the chromaticity coordinates for both the D9300 and D5000 white points are shown in Fig. 9.7. Note that each of these white points can be formed from three light-emitting elements. In one case, white is formed from red, green and blue light-emitting elements. In the other case, white is formed primarily from the cyan and red light-emitting elements with a small contribution from the green light-emitting element. However, the contribution of the green or blue light-emitting element will depend upon the location of the chromaticity coordinates of the red and cyan light-emitting elements and the selected white point. This manipulation can be performed using the methods discussed in Chap. 7 with a white mixing ratio of 1. Other colors will be formed from these colors but to understand the effect of this manipulation on stimulation of the ipRGCs, we will focus on just the formation of the display white point.

In our current application, we can then consider relying heavily on the cyan light-emitting element near our desired midday when we wish to stimulate the ipRGCs and relying more on the green and blue light-emitting elements in the evening when we do not wish to stimulate the ipRGCs. Since the cyan has been located to maximally stimulate the ipRGCs and the blue and green light emitting elements have been designed not to strongly stimulate the ipRGCs, this arrangement and drive method can significantly alter the stimulation of these cells. Table 9.9 shows the luminance of each light-emitting element and the relative normalized energy under the ipRGC sensitivity curve for each drive condition and color temperature. Note that the energy

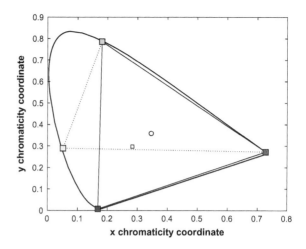

Fig. 9.7 Chromaticity diagram illustrating the chromaticity coordinates of each emitter, as well as the D9300 and D5000 white points. Note the triangles illustrate various combinations of three emitters, the smaller of which will employ a large amount of energy from the cyan emitter and the larger of which will not employ the cyan emitter

Table 9.9 Luminance of each light-emitting element (cd/m^2), assuming a 100 cd/m^2 white point, and normalized relative ipRGC energy values

Color temperature (K)	Emitter combination	Red lum.	Green luminance	Blue luminance	Cyan luminance	Relative energy
9300	RGC	31.0	7.0	0.0	62.0	1.000
	RGB	18.1	80.7	1.2	0.0	0.242
5000	RGC	30.9	33.5	0.0	35.6	0.630
	RGB	23.5	75.8	0.7	0.0	0.194

values are normalized such that the cyan-laden, 9300 K white light condition is assigned a value of 1 and the remaining values are a proportion of this value.

As shown, by relying on the RGB light-emitting elements rather than the cyan, the energy to the ipRGCs is reduced to 24% of its original value when the display white point is 9300 K. This can be compared to shifting the white point of the display to 5000 K where the relative energy is reduced to 63% of the original value when using they cyan pixel. That is this manipulation is more useful than simply shifting the white point of the display. Further, when the cyan element is not used and the white point is shifted to 5000 K, the energy available to the IPRGCs is 19% of the original value. Therefore, this display configuration can reduce the energy of the ipRGCs to 24% of the maximum value without changing the white point. Even a larger change can be observed using both emitter and white point change. As such, this multi-primary display configuration can be used to change the stimulus to the ipRGCs and possibly to affect useful biological signals. Although this example provides an interesting potential application of multi-primary displays, the true value of such a display is not certain. It is clear that illumination levels during a normal day outdoors can vary over many orders of magnitude and the color temperature of the light can also vary dramatically. As a result, this relative energy value can change by at least

9.4 Modifying Stimulation of the ipRGCs

a few orders of magnitude in our natural environment throughout any given day. Therefore, it is unclear if, or under which conditions, the relatively small change in relative energy to the ipRGCs illustrated in this application example would be significant. It is likely that the changes in color rendering, and color temperature may need to be paired with substantial changes in display luminance to have the desired magnitude of effect within this application. However, the current effect is likely on the order of the effects employed by commercial products within today's marketplace.

9.5 Enabling Display Configurations

Although the use of multi-primary display configurations can reduce power or decrease metameric failure, some display systems exist which could not be formed practically without the use of multi-primary systems. In this case, the display designer has no choice but to adopt a multi-primary display architecture. A classic example is the field sequential display. In a three-primary field sequential display, the red frame of an image is shown to the user for a brief time interval, the green frame of the image is shown to the user for a subsequent brief time, and finally the blue frame of the image is shown to the user for a third brief time. These color frames are then displayed repeatedly and rapidly to present a full color image to a user.

Unfortunately, a color artifact exists which is referred to as color breakup. This artifact occurs when an object is moved rapidly across the display or the user quickly moves their eyes with respect to the display [7]. Under these circumstances, the different color frames are registered on different areas of the human retina. Therefore, when viewing what should be a white object, the user will see red, green, and blue fringes around the edges of the object in the direction of the object or eye motion. Generally, it is believed that the eye is only able to see individual flashes if the flicker rate of a light is 60–80 Hz [4]. Therefore, one might expect that the highest possible required field rate would be somewhere between 180 and 240 Hz (i.e., 60–80 Hz for each of the three color fields). The construction of field sequential displays with field rates of 180–240 Hz is possible with reasonable power consumption. Unfortunately, the color breakup phenomenon persists at field rates much higher than 240 Hz.

> Overcoming color break-up artifacts in modern field-sequential RGB displays requires ridiculously high field rates, making these displays commercially uncompetitive.

In fact, the necessary field rate is highly dependent upon saccade velocity (i.e., the velocity of rapid eye movements between fixations). Post and colleagues demonstrated that if the user only makes slow saccades, and/or the contrast of the color edges is reduced or the luminance of the display is reduced, one can reduce the field rate

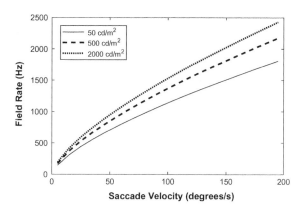

Fig. 9.8 Field rate necessary to overcome color breakup for a high contrast object (modulation of 0.95) in a typical RGB field-sequential display. Data generated from the mathematical relationship provided by Post and Colleagues [13, 14]

necessary to avoid the color breakup artifact. This relationship is shown in Fig. 9.8 [13, 14]. As this figure shows, for a modern display having a luminance of 500 cd/m^2 and a human eye saccade velocity of 100° of visual angle per second, the required field rate would need to be well above 1000 Hz. This is much higher than might be expected based upon the human eye's response to large field flickering white light. The presence of the high contrast edges in the scene significantly increases the field rate necessary for field sequential displays.

Of course, the research by Post and colleagues demonstrated that if the user only makes slow eye movements, the contrast of the color edges is reduced or the luminance of the display is reduced that one can reduce the field rate necessary to avoid the color break up artifact. Unfortunately, to be competitive modern displays require luminance values in the 500 cd/m^2 range and high contrast. Therefore, the display designer cannot practically reduce the luminance or contrast of the display and still produce a competitive display. Further, the display designer has no control over the speed at which the user moves their eyes when transferring their point of gaze from one point on the display to another. Therefore, it is not practical to manipulate any of these variables and practical displays cannot be produced with these high a field rates. Therefore, other solutions are required to permit the production of practical field-sequential displays.

Note, however, that color breakup occurs primary along high contrast edges. These edges will generally occur along the edges of white or near neutral objects or potentially along the edges of cyan or yellow objects. Color breakup does not occur along the edges of primary color objects as these edges are generally comprised of a single color and magenta edges are relatively low in luminance and therefore will often be low in contrast. Therefore, replacing a portion of the RGB luminance along these edges with white, cyan, or yellow can be effective in reducing the visibility of color breakup in field sequential displays.

The use of white fields to form a RGBW field sequential display was first discussed by Texas Instruments for use in projectors using deformable mirror devices (DMDs) [17]. These displays utilized a color filter wheel with a red, green, blue and clear

9.5 Enabling Display Configurations

segments. Light was provided by a white light source. This white light passed through the color filter wheel and was then reflected by the DMD to form a full color projection image. In this implementation, the neutral portion of each pixel was selected and this signal was used to generate a white pixel luminance for each RGB pixel. As such, neutral luminance was added to neutral pixels. This manipulation effectively decreased the modulation or contrast of the red, green, and blue fields and reduced the visibility of color breakup. While the color breakup was not eliminated, the visibility of the color breakup artifact was reduced significantly.

> The addition of broadband (e.g., white) illumination fields can significantly reduce the field rate required to yield a significant reduction in color break-up artifacts, providing commercially viable field sequential color displays.

The use of this method is illustrated in Fig. 9.9. This figure illustrates the desired image in the top left panel. As shown, in this example, the right half of the display contains a black image and the left half of the display contains a white image. Assuming the image and our eyes were stable while looking at this image, we would perceive this image appropriately whether we were viewing a pixelated color display or a field sequential display. However, in a field sequential display, artifacts can appear when either the image or our eyes move. In fact this problem is particularly problematic during high velocity saccadic eye movement, which occurs as the human eye transitions from one point of fixation to another.

To provide a conceptual understanding of this problem, we will assume that the human is fixated (c. positioned the eye fixation) on the right side of the display and begins a saccadic eye movement to the left of the display. While the eye is in motion, it continues to receive light from the environment. In a field sequential display, at the same time the eye is undergoing motion, the different fields of color are being created on the display. Therefore, during the initial portion of the eye movement, the red field might be drawn, followed by the green, and finally the blue in an RGB display. As a result, during a portion of the eye movement the visual system receives red light corresponding to the entire desired white block, as the eye continues to move the visual system then receives green, and then blue light. However, because the eye has been undergoing motion and there is a time separation between these updates, the visual system perceives the three updates as having some degree of non-overlap. Therefore, what is perceived is an image as illustrated in the top right of Fig. 9.9. Note that the red bar has a high color contrast and is very salient as is the yellow bar and the visual system detects the momentary presence of these bars, resulting in the phenomena we know as color breakup.

In Sampsell's RGBW display, half of this neutral luminance would be presented by the RGB fields and half would be presented by the W field. Therefore, the luminance of the red and yellow bars would be reduced as shown in the bottom left panel of Fig. 9.9. This reduces the contrast of these bars, making them less salient to the visual system and the magnitude and probability of perceiving this color breakup is

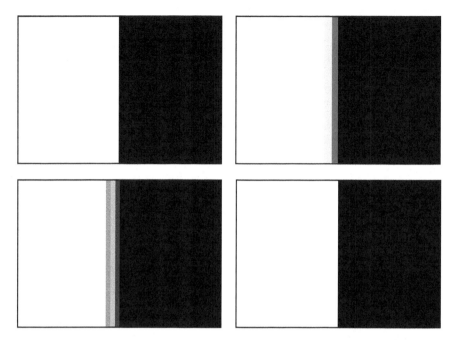

Fig. 9.9 Figure illustrating the color breakup artifacts which are likely to appear on the trailing edge of a white object during a rapid eye movement to the left. The top left panel illustrates the desired image. The top right panel demonstrates a possible stimulation on the visual system at the right edge of the white object as the eye moves from the right to the left of an RGB display. The bottom left panel illustrates the stimulation for a RGBW display where half the neutral luminance is created by the RGB primaries. Finally the bottom right panel illustrates the stimulation of the visual system for an RGBW display where all of the neutral luminance is created by the W channel

reduced. Additionally, because 4 fields a presented in the same time as the original 3 fields, the apparent size of these artifacts is also reduced, further reducing their visibility.

Finally, in an RGBW display, it is possible to move all of the neutral luminance from the RGB fields to the W fields. This eliminates the need to paint any of the white object with the RGB fields and for this neutral block, one can eliminate the artifact as shown in the bottom right of Fig. 9.9. While the artifact is eliminated for neutral-colored objects, it remains for secondary colors which require objects to be painted with two colors. However, as we have discussed elsewhere the frequency of occurrence of highly saturated, high-luminance secondary colors is low and therefore, the frequency and magnitude of this artifact is significantly reduced through this mechanism.

Approaches which include a neutral illumination portion have been shown to significantly reduce color breakup in field-sequential-displays [15]. While the field-sequential projectors explored by Sampsell used a constant white light and color filters to create the illumination source, modern versions of these displays use multiple colors of LED which can be driven to create many colors for adjustable duration within limited spatial regions of the display. As such, the flexibility of these LEDs can be leveraged to produce multiple colors of backlight, effectively creating multi-primary displays where the color of the primary is adjustable based upon image content [8, 22]. Additionally, different colors of backlight fields can be used with color filtered elements to create even more potential color combinations [9]. As such, the use of multi-primary displays or even adjustable primary displays provides the ability to reduce or eliminate color break-up in field sequential displays. As such, multi-primary displays can enable successful field sequential displays through reduction or elimination of significant image artifacts, providing an example of a display technology which is enabled by multi-primary display technology.

9.6 Summary and Questions for Thought

Previous chapters have already shown that multi-primary displays can enable expansion of the color gamut and, for high resolution displays, a reduction in the number of light-emitting elements required per pixel in a display. In this chapter we provided evidence of three additional potential advantages.

A significant advantage is improvement in power consumption. While reductions in power consumption are likely to be greater in displays which employ color filters, the examples illustrated that power savings are potentially available in emissive displays with patterned color emission. For color filtered displays, the use of multi-primary displays yield increasingly larger power advantages as the bandwidth of the color filters is reduced in displays having a white backlight. Therefore, this technology can yield displays with significantly larger color gamut while only suffering relatively modest increases in power consumption.

This chapter also illustrated that multi-primary displays can be useful in reducing metameric failure across displays, permitting colors rendered on a multi-primary display to appear more consistent across displays than would be possible otherwise. Similar to the power advantage, this advantage is greatest when the RGB primaries are narrow in bandwidth and produce a highly saturated display.

Finally, it was illustrated that multi-primary technology can be used to overcome certain display flaws, creating practical displays from technologies that would result in an unacceptable levels of artifacts or other deficiencies otherwise. This fact was illustrated using field-sequential displays. This example is relevant as more field sequential DMD projectors utilizing RGBW image fields have been produced than any other multi-primary display. Further this technology has the potential to enable the use of field sequential LCD displays, eliminating the need for LCD color filters and significantly improving the power consumption of this already dominant technology. It is worth noting however, that while such displays are likely to be employed in projection devices, liquid crystal technologies with response times fast enough to enable field sequential displays offer suffer from other artifacts, such as deficiencies in viewing angle, which reduce their utility in the direct view display market.

With this summary in mind, we can once again ask some questions to be pondered, as follows:

1. Although multi-primary displays have been demonstrated in both LCD and OLED, the only commercially available multi-primary direct view display today are the RGBW OLED displays in production by LG Display. While this technology has been demonstrated for LCD technology, why has it not received commercial adoption?
2. Although multi-primary displays have the potential to reduce the power consumption of emissive displays with patterned color, is the power savings great enough to overcome any additional manufacturing complexity?
3. Does the added advantage provided by reducing metameric failure improve the cost benefit of multi-primary patterned emissive displays?
4. Although multi-primary technology has been shown to significantly reduce or eliminate color breakup artifacts, are there other artifacts which will limit the use of this technology?
5. What other potential applications or display technologies exist which might be improved by multi-primary technology to the point that they become commercially viable?

References

1. Asano Y, Fairchild MD, Blondé L (2016) Individual colorimetric observer model. PLoS ONE 11(2):1–19. https://doi.org/10.1371/journal.pone.0145671
2. Bodner B, Robinson N, Atkins R, Daly S (2018) Correcting metameric failure of wide color gamut displays. In: Proceedings of the society for information display, Los Angeles, CA
3. Brill MH, Larimer J (2005) Avoiding on-screen metamerism in N-primary displays. J SID 13(6):509–516. https://doi.org/10.1889/1.1974003
4. Farrell JE, Benson BL, Haynie CR (1987) Predicting flicker thresholds for video display terminals. Proc SID 28(1):449–453
5. Hack M, Weaver MS, So WY, Brown JJ (2014) Novel two mask AMOLED display architecture. Digest Tech Pap SID Int Symp 45(1):567–569. https://doi.org/10.1002/j.2168-0159.2014.tb00148.x
6. Hamer JW, Arnold AD, Boroson ML, Helber MJ, Levey CI, Ludwicki JE, Van Slyke SA et al (2008) System design for a wide-color-gamut TV-sized AMOLED display. J Soc Inf Display 16(1):3–14. https://doi.org/10.1889/1.2835033
7. Johnson PV, Kim J, Banks MS (2014) The visibility of color breakup and a means to reduce it. J Vis 14(2014):1–13. https://doi.org/10.1167/14.14.10.doi
8. Kim MC (2006) Optically adjustable display color gamut in time-sequential displays using LED/Laser light sources. Displays 27(4–5):137–144. https://doi.org/10.1016/j.displa.2006.04.003
9. Langendijk EHA, Cennini G, Belik O (2012) Color-breakup evaluation of spatio-temporal color displays with two- and three-color fields. J Soc Inf Display 17(11):933–940
10. Long DL, Fairchild MD (2015) Observer metamerism models and multiprimary display systems prior experience with highly metameric color matching. Soc Motion Picture and Tel Engrs. 1100. http://doi.org/10.5594/JMI.2016.2527401
11. Lucas RJ, Peirson SN, Berson DM, Brown TM, Cooper HM, Czeisler CA, Brainard GC et al (2014) Measuring and using light in the melanopsin age. Trends Neurosci 37(1):1–9

12. Miller ME, Spindler JP (2009) OLED with magenta and green emissive layers. United States Patent Number 7,602,119
13. Post DL, Calhoun CS (1998) Predicting color breakup on field-sequential displays: part 2. In: Proceedings of the society for information display, pp 2–5
14. Post DL, Monnier P, Calhoun CS (1997) Predicting color breakup on field-sequential displays. In: SPIE head-mounted displays II conference, vol 3058, pp 57–65. http://doi.org/10.1117/12.276660
15. Qin Z, Lin YJ, Lin FC, Kuo CW, Lin CH, Sugura N, Huang YP et al (2018) Image content adaptive color breakup index for field sequential displays using a dominant visual saliency method. J Soc Inf Display 26(2):85–97
16. Ramanath R (2009) Minimizing observer metamerism in display systems. Color Res Appl 34(5):391–398. https://doi.org/10.1002/col.20523
17. Sampsell JB (1991) White light enhanced color field sequential projection. United States Patent Number 5,233,385
18. Sarkar A, Blonde' L, Le Callet P, Autrusseau F, Morvan P, Stauder J (2010) Toward reducing observer metamerism in industrial applications: colorimetric observer categories and observer classification. In: 18th color imaging conference, pp 307–313
19. Shorter PD (2015) Flashing light-evoked pupil responses in subjects with glaucoma or traumatic brain injury (Unpublished Doctoral Dissertation). The Ohio State University
20. Spindler JP, Hatwar TK, Miller ME, Arnold AD, Murdoch MJ, Kane PJ, Van Slyke SA et al (2006) System considerations for RGBW OLED displays. J Soc Inform Display 14(1):37. https://doi.org/10.1889/1.2166833
21. Yamaoka R, Sasaki T, Kataishi R, Miyairi N (2015) High-resolution OLED display with the world's lowest level of power consumption using blue/yellow tandem structure and RGBY Subpixels. In: Sid 2015 digest, pp 1027–1030
22. Zhang Y, Lin F, Langendijk EHA (2012) A field-sequential-color display with a local-primary-desaturation backlight scheme. J Soc Inf Display 19(3):55–61

Chapter 10
Virtual and Augmented Reality Displays

10.1 Defining Virtual and Augmented Reality Displays

It is first important to clearly define direct view, virtual and augmented reality displays. In this discussion, direct view displays permit us to directly view the display surface and the world in which the display resides. Every pixel on the display typically provides information to the user all of the time, filling a portion of the user's visual field of view with displayed information. Importantly, the remainder of the user's visual field is filled by the environment in which the display resides and illumination from the environment is typically reflected from the surface of the display. This environment can significantly affect visual adaptation, the contrast of the display, and provides information which is beyond the display designer's control. These displays are usually more than 5 cm in diagonal. As a result, to present an image with adequate luminance, the power consumption of these displays is typically a significant factor in their design.

VR displays are displays which are typically embedded in a headset, precluding the user from seeing any visible information other than the information shown on the display. These displays are typically imaging displays. That is they display representations of either real world scenes or an artificial scene representing a modified or artificial world. As was the case for direct view displays, every pixel in the display constantly displays information to the user. VR displays typically fill a large portion of the user's field of view, providing the user the impression that they are embedded in the world represented in the displayed image rather than the physical environment in which they are situated. To improve the perception that the user is in the alternate reality, the display is often mounted on the user's head and the displayed imagery is adjusted as a function of normal human motion, including changes in head orientation. To permit the display to be mounted on the user's head and provide a large field of view, the display is often less than 2 cm in diagonal and the image is projected through an optical path such that the image from the display fills a large portion of the human's field of view.

It is worth noting that virtual reality systems strive to stimulate the human sensory systems to cause the user to suspend the belief that they are in the world at their present place and time and replace this belief with the belief that they are in an alternate physical reality [1]. As we have discussed, as individuals we do not ever know our true physical reality but are only aware of the reality that our cognitive system forms from the sensory information that is gathered. As a result, if we can replace the stimulation provided by our natural world with alternate sensory information which provides appropriate stimulation, we should be able to create the impression of presence in the alternate reality represented by this artificial stimulation. Of course displays only stimulate the visual channel, however, this channel provides the highest bandwidth connection from our external world to our brain [2] and is therefore, very important in the formation of our impression of our world.

> The goal of VR displays is to replace the user's natural visual environment with an artificial environment which is robust enough that the user believes the artificial environment is real.

AR displays can be divided into several classes. The common attribute of these displays is that they provide information which augments or enhances the visual information in the user's real world. Like VR displays, these displays can be head-mounted and include an optical configuration to present the displayed image to the user's eyes. However, hand-held direct view displays are increasingly referred to as AR displays. This dichotomy suggests that AR displays are portable displays or displays mounted in vehicles (such as an aircraft Heads-Up Display (HUD)) which present information to augment the real world. However, enhancing the real world can be performed in one of two ways. An image of the real world can be captured and displayed to the user in near real time or the user can view the real world through a transparent surface and information can be displayed using this surface to augment the real world.

> The goal of AR displays is to provide supplemental information to the real environment, to permit the user to be more effective and efficient while performing within the real world environment. In this chapter, we restrict our discussion to head-worn AR displays.

To differentiate display technologies we will adopt a classification scheme based on display technology and presentation scheme. It should be understood that these definitions might differ from some typical classifications of these displays, which often differentiate these display types based upon the features they offer the user. However, use-based classifications can be technology agnostic and are not necessarily useful when seeking to understand the attributes of these displays.

10.1 Defining Virtual and Augmented Reality Displays

Therefore, the following definitions will be adopted in the remainder of this chapter. The term direct view displays will be reserved for displays that are viewed directly without the aid of optics which magnify the image. VR displays will refer to displays which present a full screen image, which is then magnified through optics to provide an image which fills a large portion of the user's visual field. These include displays which present a real-time image of the real world, which is captured using a real-time camera and displayed on an electronic display. The term AR displays will refer to displays which present graphics, text, or imagery which overlay a real world scene, where the real world scene is viewed through a semi-transparent surface. AR displays will often apply optics to magnify the image of the display so that it fills a large portion of the user's visual field. However, the user will look through the displayed information to see the real world.

Applying these definitions both virtual and augmented reality displays will typically be small, less than 2 cm in diagonal, and their image will be optically magnified. Because these displays are so small, the optical configuration can capture much of the light created by the display and concentrate this light into the user's eye. Therefore, unlike direct view displays, which must present an image to a large area of space, these displays tend to create a limited number of photons, a large portion of which are directed into the user's eyes. These displays typically consume much less power than the typical direct view display. In fact, in many circumstances, the display itself can consume a relatively small amount of power as compared to the display electronics. Power consumption remains a concern as the weight of the display, electronics, optics, and batteries are often supported on the head; however, the power consumption of the display itself will often be less critical than it is for direct view displays.

Power consumption in AR displays does differ somewhat from virtual reality displays. In a VR display, the display must present image information over the entire display area. However, the view of the user's external world is frequently occluded such that the only light available to the user's eyes originates within the display. As a result, the display can present light over a smaller luminance range as is typical in direct view displays. Augmented reality displays on the other hand do not typically create light for every pixel in the display but only those pixels where graphics or other content are displayed. This fact can permit displays to be formed which draw extremely small amounts of power [3]. However, the displayed content and the real world scene must both be visible to the user. When the user is viewing the real world through a transmissive surface, the display must be capable of presenting images with a very large luminance range. That is, the augmented information must not be dramatically brighter than the real world image the user is viewing if the user is going to be capable of viewing both the real world and the display. Therefore, these displays may draw extremely small amounts of power when viewing dim or dark scenes but will draw more power when viewed in bright environments.

As we define AR and VR displays in this fashion, it is important that these displays differ significantly from direct view displays along a number of dimensions. One of the more important of these dimensions to the user is that the user's "frame of reference" changes. We move through the natural world and the world provides a constant frame of reference. As such the information received by our visual system changes as we move our head and body throughout the world. However, in AR and VR displays, the information is now presented by a device which is attached to our head. Therefore, the information, by default, does not change as we move our head or walk through our world. Instead the information presented by the display to our visual system is fixed and our frame of reference changes from being world-centric to head-centric. To overcome this problem, it is necessary to update the display in response to head or body movements. Unfortunately, this correction creates the opportunity for other issues. For example, if this response is not accurate or rapid enough, a perceptual conflict will occur between the human visual and proprioceptive systems. As a result, the user is likely to experience a number of adverse physiological issues. While some of these issues are relevant to multi-primary displays and will be discussed in this chapter, the reader is referred to Patterson and colleagues for a more complete discussion [4].

10.2 Multi-primary Technology in VR/AR Displays

When discussing multi-primary displays, virtual or augmented reality displays have certain advantages. We will discuss a few of these within this section. As we have all heard, there is no free lunch, so each of these advantages have some corresponding drawbacks, which we will acknowledge.

10.2.1 Pixel Arrangements

Designers of virtual and augmented reality displays often seek to provide very high resolution images so the images will appear as realistic as they do in nature. At the same time, these displays must display images which fill a large portion of the user's field of view. Together these two requirements lead to the need to include an extremely large number of pixels in these displays. For example, Fig. 10.1 shows the number of pixels necessary as the field of view of the display is expanded, for several different pixel sizes expressed in visual angle. As this figure shows, the number of pixels necessary in the display expands nonlinearly due to the need to increase the number of pixels in both the horizontal and vertical dimensions. As we noted before, these displays are typically very small. Therefore, the pixel density must be very high. Assuming a 1.5 cm display diagonal, a 9:16 aspect ratio display, a pixel resolution of 1 min of arc and a 45° field of view, the display must contain 1.65 million pixels per cm along each dimension or 2.73 million pixels per cm^2. This can be compared to a typical desktop monitor which is designed to have about 45 pixels per cm or 2000 pixels per cm^2. Based upon this evaluation, it is clear that the pixel density in virtual and augmented reality displays is dramatically greater than they are for direct view displays and it is obvious that this fact might lead to manufacturing difficulty.

10.2 Multi-primary Technology in VR/AR Displays

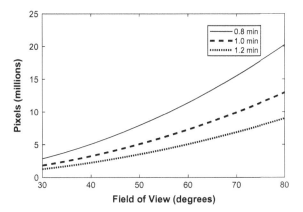

Fig. 10.1 Number of pixels as a function of pixel resolution (expressed in minutes of arc of visual angle) and field of view (expressed in visual angle)

It may not be practical to increase the number of light-emitting elements per pixel from 3 to 4 because of the extreme pixel density in these displays. After all, such a change would make a difficult manufacturing situation more challenging. We should note that the desire is to form pixels near eye-limited resolution. If this resolution is 1 min of arc or less, it is possible to adopt a pixel pattern such as the one depicted in Fig. 7.4 and reduce the number of light-emitting elements from 3 down to 2 through the use of a multi-primary display structure, reducing the number of individual light-emitting elements per cm from about 5 million to 3.3 million. In fact, in virtual and augmented reality displays, such a pixel pattern may be more likely to gain acceptance as the viewing distance to the display is typically fixed by the optical arrangement. The user cannot simply view the display from a viewing distance less than the design viewing distance to see any artifacts which might be present due to the reduction in the number of light-emitting elements as they can when approaching a direct view display. In this case, what we often consider a disadvantage of a typical multi-primary displays can become the largest advantage of this technology.

> Optics provide a known viewing distance and magnification, permitting the design of displays with alternate pixel patterns, including patterns with only 2 elements per pixel, in a display where the user cannot view the display from a distance less than the design viewing distance. This option might reduce the otherwise overwhelming display resolution requirements.

Of course, this advantage has a cost. Typically, the adoption of a multi-primary display comes with the advantage of reduced power consumption. In fact, within the actual display panel, the power consumption advantages discussed in the previous chapter will exist for Virtual and Augmented Reality displays. However, these displays require very little power to begin with and therefore the absolute power savings will be small. In fact, this power savings can be equal or even less than the power which is required to complete the image processing necessary to provide a high quality multi-primary image. The adoption of this technology might increase the overall power consumption of the display system when we include both the display panel and the electronics necessary for image processing. As a result, the application of multi-primary display technology in VR or AR displays requires efficient algorithms to minimize the power consumption during color conversion while introducing acceptable color error [5].

> Increased image processing comes with the cost of increased time necessary to complete image processing.

One other disadvantage exists which we have not discussed. This disadvantage is that the image processing required to produce the multi-primary drive signal takes time. In fact the algorithms discussed at the end of Chap. 6 require slightly more than the time required to write 1 frame of data to the display, simply because they must compute statistics over an entire frame. In most displays operating at 60 Hz, delaying the data delivery to the display by 1 frame would be inconsequential. However, in Virtual Reality displays, in particular, delaying the data by a full frame when someone turns their head within an environment has the potential to cause noticeable lag in the displayed video and can contribute to physiological effects, such as cyber sickness. Research has shown that delays in video presentation as short as 84 ms between the time the user begins a head movement and the time the image is updated are detectable by 25% of the population [6]. Therefore, it is important to maintain as short a lag as possible and this requirement may constrain the image processing techniques which are available when converting the RGB signal to a multi-primary display signal.

10.2.2 High Dynamic Range for AR Displays

In Chap. 7 we discussed high dynamic range displays as displays capable of the following three attributes; a contrast ratio greater than 10,000:1, a black level having a luminance substantially less than 1 cd/m^2, and a peak luminance level that is rendered to a luminance much larger than the luminance of diffuse white. These characteristics permit a scene to be rendered to appear very similar to the way it would appear in the natural world while the user's visual system is adapted to a normal indoor viewing environment (i.e., a visual environment having illumination levels of 200–1000 lx). However, in AR or VR displays, this assumption is often not appropriate. Instead, in virtual reality displays, the user only views the display and one would prefer to use this display to represent real world scenes. Contrarily, in augmented reality, the user is viewing the natural world which can have a very large range of luminance values and must be able to view the displayed imagery or

10.2 Multi-primary Technology in VR/AR Displays

graphics without significant disruption to their view of the environment. In either of these applications, the luminance output requirements of the display changes dramatically. In this section we will concentrate on AR displays, concentrating on VR displays in the subsequent section.

> A significant challenge in AR displays is the creation of the right amount of light to be consistent with a broad range of environmental lighting conditions.

While using AR displays, the user should be predominantly focused on the external world, using the information which the AR display overlays on the world as supporting information. As such, it is important that the AR display not occlude the external world significantly, for example by significantly reducing the contrast of the external world. Perceived contrast reductions can occur due to at least two sources.

A first source of contrast reduction is through the introduction of unintended light. As we discussed earlier, most light modulators do not block all of the light from the light source, but instead emit a small amount of light in unintended areas. This light can be particularly disruptive in AR displays. Imagine you are using your AR display near dusk. The sky has darkened but there is light enough for you to see most items even though a peak white in this environment might have a luminance less than 10 cd/m^2. As you can think of your eyes as having 3–4 log units of range, with a peak white of this value, you might see dim objects as different from black even though their luminance is less than 0.01 cd/m^2. Since you are wearing your AR display, the AR display might contain a light modulator which emits 0.2 cd/m^2 of light from any point on its surface, even when the light modulator is intended to be off, presenting a black image. If this amount of light is added to your environment, then the dimmest point in your environment is 0.21 cd/m^2 and white is now 10.2 cd/m^2. Note that the light leaking through the display reduces the available contrast range from 1000:1 to only 50:1. As such you will lose the ability to see many objects in your environment and all of the dim objects you could see without your AR display appear black. This loss of contrast has a significant impact on your ability to perform in this environment.

The second source of contrast reduction occurs as objects on the display are displayed with too high a luminance. Using the same example, we now set an icon on the display at 100 cd/m^2 even though a diffuse white in the environment has a luminance of 10 cd/m^2. In this situation the 3–4 log range of our vision is likely affected by the peak luminance in our environment so 3–4 log unit range is calculated with respect to 100 rather than 10 cd/m^2, effectively depriving us of 1 log unit of sensitivity, again causing us to lose our sensitivity to dim objects in the environment. In this situation, we likely lose sensitivity across the entire range of scene luminance values.

As such, in these displays it is critical to have control over the luminance of the display. In an LCD or other light modulator, this will mean having control of the luminance of the backlight. In an OLED or other emissive display, this will usually mean controlling the voltage range of the display drivers. Either way, this control is absolutely required. It is important to note that in a light modulator, reducing the luminance of the display reduces the luminance at every point along the tone scale, including reducing the luminance of black. Therefore, this adjustment does reduce the effect of the first source of contrast reduction even though it is easy to think of adjusting the backlight as a method of adjusting the brightest areas in the displayed image.

Note that there are two interactions which are present when we shift from an RGB to a multi-primary display. In a color filtered device, remember adding the broadband color filters provide a device that is much more efficient at transmitting light. This applies to all light. Therefore, in a light modulator, for a given power it is not only the white point that becomes brighter but the black point also becomes brighter. This is a significant problem as we are increasing both the white point and the black point of these displays proportionally when we add the more efficient color filter. However, the human eye responds nonlinearly to this light addition. The result is that the black becomes some percentage brighter as does the white point. However, the human eye responds to the contrast between these values and this increase in background luminance can produce a loss of contrast in a display which includes a light modulator. Therefore, addition of these broadband primaries can further increase the background luminance which produces a further loss of contrast.

The other effect in these multi-primary displays exists if we are using some form of dynamic luminance adjustment based upon image saturation. We now are adjusting the luminance of the display based on both the luminance of the external scene and the color saturation of the pixels in the image we are displaying. In this case, the controlling algorithms are likely to interact and this interaction must be considered.

10.2.3 Dynamic Luminance for VR Displays

Up until the last section of this book, one of the recurring themes has been the discussion of imaging systems which effectively eliminate any representation of the absolute luminance level in the original scene when displaying images. Instead, all images were discussed as being rendered to a restricted display luminance range. In fact, until recently most consumer displays produced luminance values between 1 and $100\,\text{cd/m}^2$, which is a dramatically smaller range than the 0.0001–$100{,}000\,\text{lx}$ we experience in the natural world. The use of this narrow luminance range is generally accepted as our visual system adapts to changes in absolute luminance level within our environment. Further, under low luminance conditions, our visual system's performance is much poorer than under higher luminance conditions. Therefore, the imaging industry accepts the fact that the absolute luminance level is removed from the images we are viewing.

10.2 Multi-primary Technology in VR/AR Displays

Lighting changes can occur more rapidly in the natural environment than our visual system adapts. The ability to replicate these rapid changes in lighting provide a significant challenge to creating our desired alternate reality.

While it is true that our visual system adapts to changes in luminance and we are often unaware or even incapable of sensing the absolute luminance within the range employed by displays, this adaptation is not instantaneous. In the natural world, we are very aware of dramatic changes in the light within our environment. Flashes of bright light, which occur from events such as explosions, or reductions in luminance as we walk into a dark cave, can have a dramatic influence on our emotions and our physiologic response to the environment. While the absolute luminance of the environment, once we are adapted to these environments, is not important, the radical change in lighting conditions during the transition can create a strong emotional impression. In the imaging industry, we sacrificed the ability to create this impact due to a limitation in our display technology. As this technology improves, there is no longer a need to maintain this restriction. Instead, the luminance range over which our displays will operate should increase significantly, permitting rapid changes in perceived light level to create a strong emotional response. This emotional response can be important as it can affect our ability to remember the experience.

In VR systems where the only light available to the user is the light from the display, dramatic changes in perceived luminance can be provided by controlling only the luminance range of the display. However, in more traditional display systems the ambient environment in which the display resides can have a significant effect on the luminance of the display. Unfortunately, in these environments, uncontrolled ambient light can prevent the display from having the desired impact on perceived luminance level.

The ability to control the luminance of a display can be useful to impact emotions and physiology [7]. In VR displays as the display provides the only luminance source and thus is the source for adaptation, this luminance change can potentially be dramatic, leading to an enhanced emotional response.

Dynamic adjustment of display luminance in this manner requires a change in philosophy in the imaging and display industry. Traditional image processing has assumed that the white point of the display should be fixed and the image rendered to this constant white point. The constancy of the luminance of the white point and the maintenance of luminance ratios have been assumed in practically every display produced.

When discussing the rendering of images to multi-primary displays, we discussed the fact that it can be advantageous to permit the display white point to be adjusted. In the examples we provided towards the end of Chap. 7, the ability to adjust the white point luminance of the display enables the luminance of the display to be increased substantially for most images. The original intent of the algorithms we discussed was to provide a display with as high a luminance as possible without sacrificing color saturation. As we noted in this section, the luminance could be changed rapidly at scene cuts without injecting objectionable artifacts in the video. However, the fact that these rapid adjustments are not objectionable did not imply that they are not noticeable. In fact, during evaluation, it was found that some of these changes, especially as the scene moved from low luminance to high, could introduce a startle response or have other emotional impacts on the user which they generally found desirable.

Therefore, constructing displays with constant white points or white points that change so gradually as to not be noticed by the user are not necessarily the best solution. Instead, during times of natural, rapid luminance change or when scene changes occur in which a different mood is desired, rendering the images with a rapid luminance change can be beneficial to the user's viewing experience.

Counter to applying a fully adjusted luminance, it can be desirable to control the peak luminance of the display to achieve acceptable power, lifetime, or heat within the display. To achieve these seemingly conflicting goals, it is important to consider that the human visual system is highly sensitive to rapid changes in luminance but not sensitive to slow changes in luminance. Therefore, at times when a rapid increase or decrease in luminance is desired, such a change can be implemented to elicit the appropriate emotional and physiological response from the user. The luminance of the display can then be slowly normalized to a design range so the user does not perceive this change. In this way, one could design displays to provide the desirable impact without driving substantial long term changes in power consumption or other design parameters. To achieve this goal, a method for encoding some type of absolute luminance signal should be present in the video encoding stream which serves as a master luminance control for the display. This signal can signal the desire for rapid luminance changes within the range of the display and perhaps guide the luminance of the display to provide the appropriate headroom for upcoming rapid increases or decreases in luminance.

For the display, it will be necessary to maintain appropriate luminance ratios within the scene as the luminance of the image is rapidly adjusted, permitting the display of images outside the user's current adaptation range at the points in time when these rapid adjustments occur. However, the display may then normalize the image over time to render the image with an appropriate tone scale and color for aim luminance of the display. In fact, with appropriate tone scale and color rendering, images may be rendered with lower than normal display luminance in these virtual reality environments and yet provide high quality images, as has been employed in cinema for decades. Such algorithms though must consider changes in human luminance and color sensitivity for these low light conditions [8]. These algorithms can be implemented using methods such as the rendering methods for multi-primary displays discussed in Chap. 7, dynamically changing display white point by modulating the backlight or voltage signals to the display or through some combination of these.

10.3 Barriers to VR and AR Display Adoption

Although this book is focused primarily on multi-primary displays, there are some issues which are specific to VR and AR displays which need to be overcome before wide spread adoption is likely. While these issues cannot be resolved by multi-primary technology, there are relevant interactions with multi-primary technology which will need to be considered if multi-primary displays are to be implemented in AR or VR displays.

10.3.1 Vergence and Accommodation

We discussed stereoscopic image capture in Sect. 4.4. In this section, we noted that while looking at objects in a natural scene, the human eye focuses (i.e., accommodates) to an object of interest in the natural scene and their eyes turn or converge so that the object of interest is overlaid on the fovea in each eye. We also discussed the fact that when this happens, objects which are located at different distances from the human appear at different locations on the retinas within the two eyes and that the human brain relies on this disparity in location to determine depth relative to the horopter (i.e., the plane at the vergence distance of the eyes). Further, as we discussed, accommodation distance influences the depth of field of the eye and the range of distances which are in focus at any one moment in time.

> Displays typically have a fixed focal distance while our objects in our natural environment are located at a variety of distances providing a large range of focal and vergence distances. While there are environments, such as aviation, where natural objects are located at a similar focal and vergence plane, i.e., optical infinity, these environments are the exception.

It is also important to understand that eye accommodation distance and the convergence distance are highly correlated in the natural world, the human brain can use feedback from one or both of these processes as feedback to guide the performance of the other process. Unfortunately today's stereoscopic displays, including AR and VR display technology, often does not maintain consistency of the cues associated with accommodation and convergence. Instead, commercially available displays typically are focused at a fixed distance. Therefore, the user must accommodate to the optical distance of the display to see the information that is provided. However, the convergence distance varies for objects within a stereoscopic or real world image. Therefore, these cues are not correlated when viewing these displays as they are in our natural environment.

As discussed by Patterson [9] the issue with accommodation and convergence is the following. When an individual looks at a stereo display, they can accommodate to the optical surface of the display but they can also converge to a virtual object which lies, for example, in front of the display. Presumably this situation will cause conflict—the convergence to a front position will pull accommodation off the display and a blur signal will ensue. This blur signal will drive accommodation back to the display, which in turn will pull convergence back to the display. Then the observer will again converge to look at the virtual object in front of the display, and the process will start all over again. This conundrum, commonly referred to as the accommodative-vergence conflict, can cause eye strain. However, if the depth position of the virtual image is within the depth of field of the observer's eye, then the convergence to that virtual object will pull accommodation off the display but a blur signal will not ensue because the virtual object is within the observer's depth of field. Thus, there will be no accommodative-vergence conflict causing eye strain. One might argue that AR displays are already successful in aircraft as the Head-Up Display (HUD) is popular in military and commercial aircraft and therefore, the limitations in today's displays must not be significant. Further, see-through Helmet-Mounted Displays (HMDs) have been deployed in some military aircraft, once again with the same limitations. However, in these applications, the vast majority of the real world content as well as the displayed information is placed at optical infinity. Therefore, it is rare that one needs to accommodate or converge to any distance plane other than optical infinity when viewing either the real world or the displayed information. This situation changes dramatically in automotive or other portable device applications where objects of interest can lie at distances much nearer than optical infinity. In these circumstances, the displayed information is usually rendered to some intermediate distance, for example 2 m, and the real world information is located naturally. In these circumstances the user must accommodate and converge to the distance of real world objects to view them and then accommodate to 2 m and potentially converge to some other distance to view the displayed information.

Further, while graphics can be overlaid on objects within a scene to highlight them, as is often depicted in two-dimensional renderings of augmented reality displays, natural objects exist within a three-dimensional world. Therefore, to precisely overlay graphics on real world objects in these displays, the graphics will need to be rendered with accommodation, convergence and viewing point cues which are similar to those presented by the real world object to each eye. Otherwise these graphical elements will be perceived to exist at other depth planes and may not be easily associated with the real world objects.

In VR displays, the problem is slightly less complex as objects do not have to be rendered with accommodation and convergence cues to match the cues in the real world environment. Instead, the entire image is rendered. However, in these environments, mismatches in accommodation and convergence cues can result in visual discomfort as is well known [9]. The one advantage of head-mounted virtual environment displays is that the images can be rendered to be specific to a user. Individual users are able to fuse images with different amounts of convergence both behind and in front of the accommodation plane, and it is sometimes possible to render images with different disparities for different users, especially when the distance to each object in an image is known [10]. Such systems should provide enhanced comfort, though the display cannot present images with appropriate accommodation cues.

10.3.2 Overcoming Distortion

As AR and VR displays each use optics to effectively magnify a small display, it is common for these displays to provide optical distortion of the images. Generally, the human visual system would not be sensitive to some image distortion. However, in these displays, portions of the two images which are presented to the two eyes are distorted differently. As a result, some vertical disparity can exist between the right and left eye images. The visual effects of various types of distortion have been studied and discussed in the display and human factors literature [4, 11]. Although human tolerance to distortion vary across the different types of distortion, many systems will contain multiple sources of distortion. Therefore, measures which account for multiple sources of distortion are necessary. It has been suggested that the human visual system is very sensitive to localized vertical image disparity and computing the total localized visual disparity induced by the various sources of distortion might provide a useful summary metric [12]. Interestingly, distortion can be compensated for through spatial image processing. Unfortunately, this computation is relatively complex and significant processing time can be required to rectify the image.

> When using optical systems to present information, optical and spectral distortions must be considered.

As was pointed out earlier, image information must be rendered to AR and VR displays in almost real time. Therefore, compensation of this artifact, although complex, must be performed rapidly. To meet these needs, it becomes necessary to incorporate specialized hardware, for example a field-programmable array or an application-specific integrated circuit, to accomplish this rectification.

The need to include hardware which is specialized for processing information to the AR or VR display once again provides an opportunity for multi-primary displays. If such a special hardware is required to provide image correction, could this same hardware be used to effect the RGB to multi-primary rendering algorithms, providing near real time image conversion and rendering? Further, the rectification of the image to compensate for vertical disparities or other spatial distortions created by the optics generally involves rendering the image to subpixel accuracy and this process is likely to interact with any rendering algorithms which render images to multi-primary pixel structures.

10.3.3 Foveated Imaging

To provide a completely immersive experience, in which the user feels as though they are present in the virtual world, large fields of view, perhaps as large as 140° of visual angle for a direct view display, are required [13]. However, the time to process and transmit an image with this large a field of view at near eye limited resolution can be excessive. One approach to overcoming this problem is to take advantage of the spatial characteristics of the human eye.

As we discussed earlier, the fovea is a portion of our eye which contains the preponderance of cones. This portion of our eye covers an area about 2° of visual angle. Outside of our fovea, the resolution of our eye drops continuously throughout the remainder of our visual field as the distance into our periphery increases. Therefore, the resolution of our eye is much less at 10° from the center of our fovea than it is in the center of our fovea and this resolution continues to decrease as the distance into our periphery increases. This fall-off has been characterized through the use of peripherally-dependent contrast threshold functions [14–16]. Therefore, if our eyes never moved, we could render all images to be high resolution in the center and much lower resolution in our periphery. Since, we move both our eyes and our head as we explore every environment, constantly sampling our environment with high resolution snapshots, a static rendering of ever decreasing resolution from the center of the display is not practical and users will quickly see the loss of resolution outside the center of the display.

It has been demonstrated that if the images can be updated rapidly enough and if the location of the fovea within the image can be reliably determined, one can render images with lower peripheral resolution in a way that the degraded information is not visible [17]. In this particular example, a laboratory environment was used to exactly determine the fixation point of the user and efforts were made to match to the resolution of the displayed images to the user's capability. The research demonstrated that as long as modulation was maintained, anytime the modulation was greater than the peripherally dependent contrast threshold function, the image degradation was imperceptible. However, once modulation was removed which was above the peripherally-dependent contrast threshold function, the degradation was detectable.

This particular implementation is likely not practical for use in commercial displays. However, significant advantages could be obtained for much less aggressive rendering techniques. For instance, it is unlikely that users will make eye movements larger than about 30°. Therefore, one might render as much as the 60° in the area of the user's current fixation with high resolution and only reduce the resolution beyond this excessively large eye box. The likelihood of the user seeing errors with such a system is small and the need for precise knowledge of eye fixation location and instantaneous image update is reduced. In a display with a 140° field of view, 80° of this field of view can be rendered with significantly reduced resolution, reducing the image bandwidth by a large margin.

At this time, however, the implications of foveated imaging for multi-primary displays remain unexplored. It is certainly true that the vast majority of the cones are located in the fovea and color sensitivity decreases in the periphery. Therefore, it is possible that color information can be reduced in the periphery, permitting the multi-primary rendering algorithms to rely even heavier on the additional light-emitting elements used to create the neutral content in the image than on the saturated color primaries. This effect and the need for low resolution rendering in the periphery might provide an opportunity for future wide field of view VR displays.

10.3.4 Cyber Sickness

Also tied to the need for rapid image processing in VR systems is the phenomena of cyber or simulator sickness. In this phenomena individuals experience symptoms similar to motion sickness, including nausea. A contributing factor to this phenomena arises as the human brain uses multiple sensory systems to sense self-motion [18]. One set of sensors is in the inner ear which sense acceleration. Simultaneously, our skin and the proprioceptive system senses air flow and pressure resulting from acceleration, leading to the sensation of motion. Of course our visual system is also used to assess self-motion.

Individuals are usually stationary while using a VR system. Therefore, the otoliths in the inner ear and the proprioceptive system provide signals to the brain consistent with the fact that the individual is stationary. At the same time, while watching motion video on a wide field of view VR display captured with a camera which is moved through the environment, the visual system is providing a strong signal that the individual is moving. As these two signals are in agreement while operating in the natural environment, the human brain has difficulty forming a consistent perception from these conflicting signals. The result is often the physiological response we know as cyber sickness.

> Cyber sickness results from an array of system interactions, and the effects of this phenomenon can be debilitating. Control or elimination of this phenomenon is crucial to market adoption of VR displays.

Research shows that both foveal and peripheral vision play key roles in the perception of visually-induced self-motion, often referred to as vection [19]. Further, wide field of view displays are typically associated with an increase in the perception of self-motion [13]. Therefore, the user is most susceptible to cyber sickness when motion imagery is shown. This is particularly true when there are multiple visual cues to acceleration and deceleration on a wide field of view, which precludes any view of the stationary environment in which the person resides. However, cyber sickness is influenced by many other factors including the length and frequency of exposure [20] and the degree of user control within the virtual environment [21].

Similar sensations can be introduced by other effects. For instance, adding latency to a system in response to head movements can also create conflicts between the vestibular system in our inner ear and our visual system, leading to cyber sickness [22]. Under these circumstances, the user moves their head and both the vestibular system and the kinesthetic system, which responds to muscle movements, signal that the head has undergone motion. However, when it takes the VR system a perceptible period of time to sense this motion and to render the images appropriate with head motion, the sensation of motion produced by the visual system is delayed compared to the sensation which occurs naturally. Once again, this mismatch between the different sensations provided to the brain results in conflict and can produce cyber sickness. Recent research has shown that lengthening a constant delay in the head movement to display rendering loop does not increase the onset of cyber sickness [23], perhaps because people are able to rapidly adapt to a constant delay. However, introduction of a variable delay is more likely to cause the onset of cyber sickness [24] and offsets in the frequency of the update and the information which is displayed might also increase the likelihood of cyber sickness [25]. Further, one must insure that the perception of the size of head movements and the physical movement of the head is maintained within a VR display to avoid cyber sickness [22]. As a result, although delays in updating the display in response to head movement might be noticeable to a user, as long as these delays are constant, they may not increase the likelihood of cyber sickness. Therefore, any delays necessary for converting RGB to multi-primary image content will likely not increase the likelihood of cyber sickness, as long as these delays are constant.

10.4 Summary and Questions for Thought

Although VR and AR displays are a small subset of the display market today, VR displays provide an interesting market for multi-primary displays. They offer both: (1) the potential to reduce the number of light-emitting elements required to provide eye limited resolution and (2) the potential to provide dynamic luminance adjustment. These are two difficult to achieve, but desirable, attributes of VR displays. The processing time necessary for color conversion and subpixel rendering provides a significant challenge, due to the need to provide rapid updates in response to head movements. Unfortunately, multi-primary display technology does not provide clear methods to address accommodation-vergence mismatch or cyber sickness, two of the likely barriers to broad adoption of VR and AR displays.

With this discussion, here are a few questions for thought:

(1) When are multi-primary displays advantageous to RGB displays for VR displays?
(2) Is there ever and advantage of multi-primary displays for AR displays?
(3) Is there any advantage of constructing VR displays with different densities of light-emitting elements at different distances from the center of the display? If

so, should we also consider changing the number of additional primary light-emitting elements with respect to the number of RGB light-emitting elements as a function of the distance from the center of the display?
(4) How might display attributes such as increased display luminance, expanded color gamut, or higher resolution imagery really affect cybersickness?

References

1. Sheridan TB (1992) Musings on Telepresence and virtual presence. Presence Teleoper Virtual Environ 1(1):120–126. http://doi.org/10.1162/pres.1992.1.1.120
2. Furness TA, Barfield W (1995) Introduction to virtual environments and advanced interface design. In: Furness TA, Barfield W (eds) Virtual environments and advanced interface design. Oxford University Press, New York, pp 3–13
3. Vogel U, Beyer B, Schober M, Wartenberg P, Brenner S, Bunk G, Richter B (2017) Ultra-low power OLED microdisplay for extended battery life in NTE displays. Soc Inf Disp Techn Papers 48(1):1125–1128
4. Patterson RE, Winterbottom MD, Pierce BJ (2006) Perceptual issues in the use of head-mounted visual displays. Hum Factors 48(3):555–573. https://doi.org/10.1518/001872006778606877
5. Can C, Underwood I (2013) Compact and efficient RGB to RGBW data conversion method and its application in OLED microdisplays. J Soc Inform Display 21(3):109–119. https://doi.org/10.1002/jsid.158
6. Moss JD, Muth ER, Tyrrell RA, Stephens BR (2010) Perceptual thresholds for display lag in a real visual environment are not affected by field of view or psychophysical technique. Displays 31(3):143–149. https://doi.org/10.1016/j.displa.2010.04.002
7. Brunick KL, Cutting JE, Delong JE (2013) Low-level features of film: what they are and why we would be lost without them. In: Shimamura AP (ed) Psychocinematics: exploring cognition in movies. Oxford University Press, New York, NY, pp 133–148
8. Mantiuk R, Myszkowski K, Seidel H-P (2006) A perceptual framework for contrast processing of high dynamic range images. ACM Trans Appl Percept 3(3):286–308
9. Patterson RE (2015) Human factors of stereoscopic 3D displays. Springer, London, UK
10. Jin EW, Miller ME, Endrikhovski S, Cerosaletti CD (2005) Creating a comfortable stereoscopic viewing experience: effects of viewing distance and field of view on fusional range. In AJ Woods, MT Bolas, JO Meritt, IE McDowall (Eds), Society of photo-optical and instrumentation engineers proceedings: stereoscopic displays and virtual reality systems XII, vol 5664, pp 5612–5664. http://dx.doi.org/10.1117/12.585992
11. Rash CE, Russo MB, Tetowski TR, Schmeisser ET (eds) (2009) Helmet-mounted displays: sensation, perception and cognition issues. U.S. Army Aeromedical Research Laboratory, Fort Rucker, AL
12. Jin EW, Miller ME (2006) Tolerance of misalignment in stereoscopic systems. In: Proceedings of the international congress of imaging science (ICIS) '06. Society for Imaging Science and Technology, Rochester, NY, pp 370–373
13. Lin JJ-W, Duh HBL, Parker DE, Abi-Rached H, Furness TA (2002) Effects of field of view on presence, enjoyment, memory, and simulator sickness in a virtual environment. Proc IEEE Virtual Reality 2002(January):164–171. https://doi.org/10.1109/VR.2002.996519
14. Anderson SJ, Mullen KT, Hess RF (1991) Human peripheral spatial resolution for achromatic and chromatic stimuli: limits imposed by optical and retinal factors. J Physiol 442:47–64
15. Banks MS, Sekuler AB, Anderson SJ (1991) Peripheral spatial vision: limits imposed by optics, photoreceptors, and receptor pooling. J Opt Soc Am A 8(11):1775–1787. https://doi.org/10.1364/JOSAA.8.001775

16. Peli E, Geri GA (2001) Discrimination of wide-field images as a test of a peripheral-vision model. J Opt Soc Am A Opt Image Sci Vis 18(2):294–301
17. Loschky L, McConkie G, Yang J, Miller M (2005) The limits of visual resolution in natural scene viewing. Vis Cogn 12(6):1057–1092. https://doi.org/10.1080/13506280444000652
18. Coren S, Porac C, Ward LM (1984) Sensation and perception (Second). Academic Press Inc, Orlando, FL
19. Wolpert L (1987) Field of view versus retinal region in the perception of self motion. Dissertation, The Ohio State University, Colombus, OH
20. Kennedy RS, Stanney KM, Dunlap WP (2000) Duration and exposure to virtual environments: sickness curves during and across sessions. Presence 9(5):463–472
21. Wampold BEB, Brown GS (Jeb) J. G. K. G. S., Pieterse AL, Lee MMY, Ritmeester A, Collins NM, Fishman DB (1999) Copyright © 1999. All rights reserved. Int J Lang Commun Disord/Royal Coll Speech Lang Ther 42(2):130–153. http://doi.org/10.1086/250095
22. Palmisano S, Mursic R, Kim J (2017) Vection and cybersickness generated by head-and-display motion in the Oculus Rift. Displays 46:1–8. https://doi.org/10.1016/j.displa.2016.11.001
23. Moss JD, Austin J, Salley J, Coats J, Williams K, Muth ER (2011) The effects of display delay on simulator sickness. Displays 32(4):159–168. https://doi.org/10.1016/j.displa.2011.05.010
24. St. Pierre ME, Banerjee S, Hoover AW, Muth ER (2015) The effects of 0.2 Hz varying latency with 20–100 ms varying amplitude on simulator sickness in a helmet mounted display. Displays 36:1–8. http://doi.org/10.1016/j.displa.2014.10.005
25. Groen EL, Bos JE (2008) Simulator sickness depends on frequency of the simulator motion mismatch: an observation. Presence Teleoper Virtual Environ 17(6):584–593. http://doi.org/10.1162/pres.17.6.584

Chapter 11
Multi-primary Displays, Future or Failure?

We have discussed image systems throughout this book, leading to a discussion of multi-primary displays. As we have discussed, there are numerous potential advantages of including more than the standard three, i.e., red, green, and blue, primaries in a display. We have discussed improvements in power efficiency, reductions in observer metamerism, expanded dynamic range, improved perceived spatial resolution, potential biorhythm control, and enabling display technologies which would have significant market deficiencies otherwise. While the benefits discussed are all significant, one might argue that they often do not provide as dramatic an improvement in user value as did the addition of color, popularized in the 1950s; the adoption of higher resolution displays, popularized through the adoption of high definition television in the 2000s; or the adoption of wide field-of-view displays, popularized through the adoption of large flat screen LCDs in the 2010s.

Multi-primary displays have been discussed since the 1980s [5]. Since that time numerous prototype multi-primary displays have been demonstrated and research has continued to expand in this area. In fact, according to Google Scholar^{tm} [2] the Silverstein and Monte patent has been cited 205 times. Among these citations, 194 occurred in the year 2000 or after and 112 occurred in the year 2010 or after. Despite this interest, to date there have been only two commercially successful display examples. The first of these was Texas Instrument's Deformable Mirror Device Projectors, which were commercialized in the 1990s. Inclusion of a white channel in these displays reduced color breakup artifacts to an acceptable level and improved the luminance efficacy of these devices. The more recent example is LG's RGBW OLED televisions. The use of a white emitter with color filters made manufacturing of these displays possible and the RGBW color filter arrangement reduced the power consumption of the display sufficiently to make the display competitive within the market place. Perhaps inclusion of the additional primary permitted each of these display technologies to become commercially successful. Yet the vast majority of displays produced today utilize the standard RGB format.

Arguably, inclusion of an additional primary in today's displays adds complexity and risk during manufacturing. Without providing clear market advantages, this additional risk is simply not justified. In this chapter, we will discuss the future of multi-primary display technology, with the goal of understanding whether this technology is likely to be adopted more frequently in the future or if the application of this technology will continue to be an anomaly.

11.1 Reviewing the System

Without an understanding of the environment and system in which the display resides, the advantages of multi-primary displays are not fully evident. In an imaging system, we can begin to understand the utility of multi-primary technology by understanding natural scenes. These scenes are composed of an array of objects which reflect different bands of wavelengths of light which our brain sees as colors. Most of these objects reflect a broad bandwidth or multiple bands of wavelengths, resulting in the perception of relatively desaturated color. Therefore, the reproduction of these colors does not require highly saturated primaries. Although rare, objects exist which reflect narrow bandwidths of wavelengths, resulting in the perception of highly saturated colors. Because these objects are rare, their perceived color tends to have a higher impact on human perception. As such, reproduction of these rare, highly-saturated, colors is important in modern displays. When saturated colors are present in scenes, they are often low in luminance compared to other colors. Therefore, the need to reproduce high luminance, highly-saturated colors is extremely rare.

As a result, we can consider image content among the central drivers for multi-primary displays. Figure 11.1 shows the relationships between many of the concepts related to multi-primary displays. As shown, content, whether it is images or graphics, is near the center of this diagram. This content affects the saturation that is required and drives the need to use narrowband RGB primaries or broadband additional light-emitting elements in a multi-primary display.

In this figure, influences generally flow from the top to the bottom. Therefore, RGB color is created through one of two general approaches. These include spatial methods, in which differently-colored light-emitting elements are arranged side by side, or temporally, in which subsequent frames of information contain different colors. Although not discussed to a great extent, one could consider displays which include some combination of spatial and temporal color formation as well. While the typical flow in this diagram is from top to bottom. Any flows which reverse this direction, for example the influence of adding an additional color which influences items above it in the diagram, are depicted by connectors with arrow heads indicating this reversal of direction.

As you look at Fig. 11.1, you will notice the use of color to indicate various items. Content has its own color. Color types are depicted in yellow. System design decisions or parameters are indicated in gray. Finally the attributes of the displays that are being influenced are depicted in green. An important attribute of this diagram is that it does not depict a linear flow. Instead the diagram indicates loops, which indicate interactions that should be considered during display design. While this

11.1 Reviewing the System

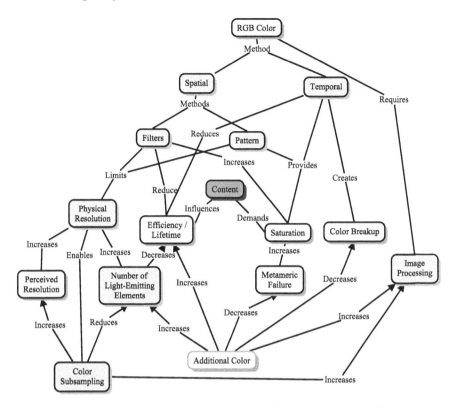

Fig. 11.1 Concept map depicting the inter-relationships related to multi-primary displays

diagram does not consider all of the important attributes of the display or their effects, it provides a high level overview of many of the effects we have discussed in this text.

Walking through this figure further, we can see that spatial methods for forming color generally consist of displays using color filters or patterned materials, although, once again, displays could be formed using both approaches. Both the presence of color filters and the other methods of forming color generally will reduce either the efficiency, lifetime or both of the display. This occurs as less time or area is available to form each color in temporal or patterned displays. The effect is often larger in displays using color filters as these displays often suffer not only these effects, but the color filters absorb an additional portion of the light. Of course, the efficiency and lifetime are often influenced by the image content as well. As the efficiency improves, the overall efficacy of the display (i.e., the ratio of luminance to power consumed) also increases. However, efficacy can also be improved by adding additional colors in the display as these colors can contain more energy in the portions of the visible spectrum where our eye's sensitivity is greatest.

Spatial methods each limit the physical resolution of the display. As the physical resolution increases perceived resolution also increases. As we discussed, this perceived resolution can also be improved through the use of color subsampling when the resolution of the display is high enough. That is having a high physical resolution enables color subsampling. Color subsampling and resolution each influence the number of light-emitting elements, which also affects efficiency or lifetime as less area is available for light emission, transmission, or reflection in patterned displays as the number of light-emitting elements increase. The number of light-emitting elements is also certainly influenced by the number of additional colors the display must produce.

Many of the design decisions which are made in displays affect the maximum color saturation available in a display. Increases in saturation are generally valued in the display industry. However, as saturation increases, the likelihood and the magnitude of metameric failure also increases. Fortunately, the use of broad bandwidth emitters or an increased number of emitters can be used in conjunction with the narrow bandwidth emitters to provide the potential for exceptional color saturation as well as to reduce the likelihood and magnitude of metameric failure.

Although most of our attention was given to displays in which color was formed through spatial patterning, we also mentioned displays with field sequential color. As the diagram shows, saturation can be provided through this technique as well. However, temporal approaches to color often suffer artifacts, such as color breakup. Once again, judicious selection of additional color primaries can help to overcome this artifact. Finally, we see that to form a high quality RGB image, image processing is required. The amount of image processing is increased as the RGB signal is converted to include signals for additional primaries or when processing is needed to support spatial interpolation to alternate pixel patterns or the use of expanded luminance range.

Each of the attributes of these system components are important to understand the value of multi-primary displays. Without this understanding, the argument for multi-primary displays is weakened.

11.2 Challenges to Multi-primary Displays

There are a number of challenges which make the adoption of multi-primary displays difficult. The predominant challenge is manufacturing. To date, the addition of primaries has required either the designation of a time slice in field sequential displays or additional light-emitting elements in spatially patterned display structures. Each of these add both design and manufacturing complexity. Additional resources must be spent to develop algorithms for color conversion and spatial rendering.

From a manufacturing perspective, additional primaries in spatially-patterned displays have in the past required additional manufacturing steps to form each colored light-emitting element. Moreover, the adoption of an additional primary has implied an increase in the number of light-emitting elements per pixel. These additional

11.2 Challenges to Multi-primary Displays

manufacturing steps and the increased manufacturing tolerances implied by adding light-emitting elements have increased the probability of artifacts in these displays, implying a reduction in display yield (i.e., the number of marketable displays as a proportion of the number of displays manufactured). This reduction in yield implies an increase in manufacturing cost and a reduction in display profitability. Obviously, this risk is not a risk that corporations are likely to shoulder without the potential for significant reward.

From a design perspective, new algorithms are required. These algorithms are, at times, specific to the particular display implementation. Subpixel rendering, for example, must be tuned to each particular display. The algorithm itself may not be novel to each display but it must be tuned to each display resolution. Further, the color conversion algorithms have not been well understood. However, with the investment and advances in these algorithms over the past couple decades, this problem is becoming less of a concern. Nonetheless, it remains a concern that it is difficult to understand the image quality of these displays before they are constructed. Additional research in this area may be beneficial to the adoption of this technology.

There are additional difficulties. For example, display panels are often manufactured by a different company than the processing chips used to provide images to these display panels. Further, each of these companies are often separate from the display integrator who packages these components into a marketable display. In many markets, the panel and processor chip manufacturers consider the display integrator the customer and develop products based on requirements provided by the integrator. Unfortunately, the integrator often does not have the technical knowledge to understand the potential advantages of multi-primary display technology. As such, they are not likely to engage in technology pull. Under these circumstances, the display panels and the image processors are each being optimized assuming an RGB architecture, without considering the larger system-wide optimization of the final display system.

> This feedback loop from the customer to the supplier is a tenant of "Total Quality Control", a concept which has been shown to be an important tenant in product design and manufacturing. As stated by Dr. Ishikawa, "the manufacturer must always be keenly attentive to consumer requirements, and the opinions of consumers must be anticipated as the manufacturer establishes his own standards. Unless this is done, Quality Control cannot achieve its goals, nor can it assure quality to consumers." [3]

Customers provide the final challenge to the adoption of multi-primary displays. Consumers may not care specifically about this technology. Instead, they are focused on the resulting display attributes. However, there is a certain resistance to change, where individuals are cautious of changes in technology. Education of the sales force and the customer requires time and money, each of which a display provider may wish to spend on alternate activities. Further, while customers may see advantages of this technology when viewing the displays at the design viewing distance, they often view displays from other distances. For example, manufacturers have fielded RGB products in recent years which contained more green light-emitting elements than red and blue or more red and green light-emitting elements than blue. These display designs take advantage of the fact that the green light-emitting element carries the majority of the luminance information and that the human visual system relies primarily on luminance for high resolution information. Therefore, a display containing more high luminance elements can provide a higher perceived resolution than a display having an equal number of light-emitting elements but equal numbers of high luminance green and low luminance red and blue light emitting elements [4]. Unfortunately competitors have encouraged display measurement companies to highlight the color uniformity deficiencies that exist when these displays are viewed at a distance closer than the design viewing distance. The resulting negative press coverage has limited the adoption of this technology. Therefore, it is necessary to develop and clearly communicate a message which highlights the advantage of technological changes so that customers do not become overly fixated on the negative attributes of the tradeoffs made during technology development. Once again, however, developing this marketing message requires time and money.

The final challenge is simply the fact that any change towards multi-primary display technology is exactly that; change. Change implies risk and the appetite for risk in a commodity market is small. In the end, the question is: "What is required to make the risk of adopting multi-primary technology worth the potential reward?"

11.3 Supporting Trends

Although there are significant barriers to the adoption of multi-primary displays, there are significant trends which provide long term support for this technology. The three most significant trends are the drive towards larger color gamut, higher resolution, and high dynamic range. In general, the advantages of multi-primary displays increase as a function of increases in each of these display attributes.

We can observe the drive to larger color gamuts through multiple advances. First, advances in modern image encoding standards have adopted larger and larger color gamuts. Second, display manufacturers are adopting technologies, such as iLED and quantum dot emitters, which enable larger color gamuts. With increases in larger color gamuts, the emitters become more pure, increasing the likelihood of metameric failure. Further, in displays using color filters, this drive encourages the use of narrower bandwidth color filters, which reduce the energy efficiency of each primary. As we discussed in Chap. 9, the adoption of multi-primary display technology can help to overcome each of these issues while still providing the desired increases in color gamut. Therefore, each of these trends are likely to improve the value of multi-primary technology within the display industry.

The display industry continues to push towards higher resolution displays. Marketing of HDTV in the 2000s was very successful in aiding the consumer replacement of early LCDs as consumers reacted positively to the availability of higher resolution displays. More recently, Apple Inc. successfully marketed the Retina™ Display as a higher resolution display for computers and mobile devices. This marketing campaign permitted the company to gain market share as the increased resolution improved the value these displays offered consumers. This success has led to the adoption of higher and higher resolution devices. Prior to 2010, displays having 4 K resolution were unheard of outside the Cinema projection and high-end display markets. Now, displays having 8 K resolution are commonly discussed, many video streaming companies offer content with 4 K addressable pixels, and 4 K consumer displays are available for use as desktop monitors and televisions.

The traditional application of multi-primary displays where three light-emitting elements are replaced with four or more light-emitting elements to display the content associated with each pixel might imply that the trend towards higher resolutions displays might preclude the adoption of multi-primary displays. After all, the manufacture of 4 K or 8 K displays with three colored elements per pixel is difficult and complex. Adding one or more additional colored elements per pixel only increases this complexity. However, as we discussed in Chap. 8, once the number of addressable pixels reach or exceed 4 K, the human's ability to see artifacts created by defining a pixel as a luminance bearing element with lower resolution color information is virtually eliminated at viewing distances in excess of 3 picture heights. Therefore, in high resolution, multi-primary displays having four differently colored light-emitting elements, it is possible to create displays having only two elements per pixel without introducing visible artifacts. This is especially true for displays which have significantly more than 4 K addressable pixels. It is reasonable to adopt multi-primary 8 K technology having two elements per pixel for a total of 16 K light-emitting elements as an intermediate between the adoption of RGB technology with 4 K and 8 K addressability, which requires an increase from 12 K to 24 K light-emitting elements, respectively. In fact, the utility of an 8 K display having 24 K light-emitting elements is questionable, especially if the display has a large color gamut, unless the user is likely to view the display from a distance much less than 3 picture heights. As we noted in Chap. 10, this tradeoff may be particularly advantaged in VR displays, where the display viewing distance and magnification is fixed.

The final important trend involves the adoption of High Dynamic Range (HDR) displays. It is clear that interest in HDR displays is increasing. Samsung has recently began marketing their QLED line of LCDs. This technology permits control of the LEDs which provide illumination to different spatial areas within the backlight. Control of the light output by different spatial areas within the backlight, independent of the LCD, permits one to create a HDR display. Similarly, LG has introduced RGBW OLED TVs, which are advertised as HDR displays. At the same time, various image and video storage standards are under development to enable the sharing of HDR content. Each of these trends illustrate the industry's interest in HDR.

In traditional displays, the luminance range of the display is limited so that an image can be displayed without artifacts using 8 bits of luminance information. Therefore, today's displays utilize 8 bit drivers which convert input code values to voltage inside the LCD or OLED display. The display of HDR information, however, has the potential to significantly enhance the range of luminance information and the industry has discussed the need for additional bit depth in the imaging pipeline as well as in display drivers to avoid false contours or other artifacts [1]. Although one could apply dithering or other methods to obscure these artifacts, multi-primary display technology can be advantageous in this area as well. As we have alluded, the luminance range of the additional primaries in multi-primary displays can be designed to be significantly larger than the luminance range provided by RGB. Therefore, one can use the additional primaries to create the larger luminance range required for HDR displays and use the RGB values to render luminance values between those that can be provided by the additional primaries. As such, the redundancy provided in these systems can be used to increase the perceived bit depth while applying only 8 bit drivers within the display. Such a rendering approach provides one more advantage for multi-primary displays.

One final trend must be discussed. Remembering back to Chap. 1, we mentioned a few applications, including cyber shopping and medicine, where color accuracy is important to the business case of these potential markets. Color accuracy has typically not been the focus in display development. Instead the industry often focused on providing preferred color, which often included distortions in color rendering to provide a display with brighter, more saturated colors, than existed in the original scene. This work extended into multi-primary displays where many of the color conversion algorithms purposefully distorted color, usually to provide brighter or lower power images. These distortions are certainly not necessary and therefore the color conversion algorithms discussed in this text sought to maintain a metameric match to the original color palette. Therefore, it is certainly realistic to construct multi-primary displays which are capable of providing either accurate or preferred color, depending upon the user's or the application's needs. However, to achieve this, one must develop an image processing path capable of providing accurate color and then purposefully distort this color when providing preferred color. The use of multi-primary displays certainly does not preclude one from providing accurate color. In fact, multi-primary displays can be advantageous in this area as they help to overcome metameric failure.

As we have discussed, multi-primary displays provide the opportunity to increase the color gamut of the display while limiting the risk of metameric failure, reducing the number of elements in very high resolution displays, and reducing the required bit depth for drivers within a HDR display. Each of these attributes is significant given current display-industry trends. Therefore these trends support the continued use and adoption of multi-primary displays, perhaps justifying the increased interest in this category of displays as evidenced by the increasing number of citations to Silverstein and Monty's original patent within this technology area. As such, the legacy of Lou Silverstein's and Robert Monty's early work in multi-primary display would appear to have a bright future.

References

1. Daly S, Feng XF (2005) Bit-depth extension: overcoming LCD-driver limitations by using models of the equivalent input noise of the visual system. J Soc Inf Disp 13(1):51–66. https://doi.org/10.1889/1.1867100
2. Google Scholar (2018) https://scholar.google.com/scholar?cluster=16717512251991780440&hl=en&as_sdt=5,36&sciodt=0,36. Accessed from 21 Apr 2018
3. Ishikawa K (1985) What is total quality control? the Japanese way. Prentice Hall, New York, NY
4. Kranz JH, Silverstein LD (1990) Color matrix display image quality: the effects of luminance and spatial sampling. In: SID symposium digest of technical papers, pp 29–32
5. Silverstein LD, Monty RW (1989) Four color repetitive sequence matrix array for flat panel displays. United States Patent Number 4,800,375

Glossary

Accommodation the focusing of our eye at a distance associated with a display or an object in a natural environment.

Additional Color any colors of light-emitting elements in addition to red, green, and blue which is employed within a multi-primary display.

Anode a positively charged electrode through which current flows into a device.

Aperture a space through which light passes to a sensor in a camera, reducing the aperture reduces the amount of light which reaches the sensor while increasing the aperture increases the amount of light reaching the sensor.

Augmented Reality a term used to describe systems in which graphics or other information is overlaid on the real world to improve user performance or understanding. Although many display technologies could be applied to provide this feature, within this text augmented reality displays refer to displays which project an image of a display onto a transparent surface between the user and the world.

AR Augmented Reality (AR)

Blackbody an idealized physical object which absorbs all electromagnetic energy which enters it. To maintain thermal equilibrium a blackbody must emit electronic radiation and the spectrum of this radiation is determined by the temperature of the blackbody as described by Planck's Law.

Blackbody Locus a curve within the chromaticity diagram indicating the chromaticity coordinates a blackbody undergoes as it is heated.

Brightness the perception elicited by the luminance of an object in the visual environment.

Cathode a negatively charged electrode through which current flows out of a device.

CCT Correlated Color Temperature (CCT)

Color Breakup a visual artifact present in field sequential displays where artifacts appear on the edges of objects undergoing motion or on the edges of objects as the user moves their eyes across the display.

Convolve a mathematical process involving two lists or arrays of numbers in which each element in one list is multiplied by the corresponding element in the second list resulting in a single array of numbers. This array of numbers is often summed to provide a single value, for example, the total amount of energy passing through a color filter.

Color Temperature an indication of the color of a light which has chromaticity coordinates on or very near the blackbody locus. The color temperature indicates the temperature of a blackbody which achieves equivalent chromaticity coordinates.

Color Rendering Index a metric used to judge the quality of a lamp with respect to a reference illumination. Lamps with higher CRI are more likely to illuminate an object such that its color appears the same as it would if illuminated the reference illumination.

Correlated Color Temperature a measure of the color of a white light source. The correlated color temperature indicates the nearest iso-temperature line to the chromaticity coordinates of a light source, where iso-temperature lines are lines drawn perpendicular to the blackbody locus within the chromaticity diagram.

Cyber Sickness a phenomena experienced by individuals while using virtual reality displays which results in symptoms similar to motion sickness.

Depth of Field a range of distances over which an image is in focus within an imaging system. Depth of field is generally increased by decreasing the aperture in a photographic system.

Down-Conversion a physical process in which a particle (often a photon) having an energy excites a molecule. The molecule then undergoes relaxation giving off energy and a lower energy particle (photon). This process is useful as it converts high energy, short wavelength photons to lower energy, longer wavelength photons (i.e., converts blue to red or green light).

Dynamic Range the ratio between the maximum and minimum light level that can be measured by an instrument or perceived by the human eye.

Efficacy a measure of the luminous efficiency of a light source. This value is obtained by convolving the power output by a lamp by the eye's sensitivity function and dividing the total weighted power by the electric power necessary to create the light output. Because this quantity is weighted by the eye's sensitivity function, the efficiency of a light in producing light at 560 nm typically influences the result more than the efficiency of the light near the extremes of the visible spectrum.

Efficiency a measure of the total optical power created by a light source divided by the electric power necessary to create the optical power.

Fabry-Perot Microcavity a microcavity formed using two highly reflecting mirrors to create a standing wave having a sharp resonance, thus producing very narrow-band energy emission.

Focal Plane the plane at which a lens or optical system focuses the light from a real image.

Field of View the angle subtended on the user's retina by the display.

Flare the spreading or diffusion of light as it passes through or between layers of optical media.

Fovea the small area on the human retina which has the highest density of cones, providing the ability to resolve high spatial frequency information.

Gamut the geometric entity formed by connecting the most extreme chromaticity coordinates of a family of display primaries within a defined color space. For example, connecting the chromaticity coordinates of red, green, and blue light-emitting elements forms a triangle, which is the gamut of a typical tricolor display.

Glare the perception of unwanted light which occurs as a result of flare within an optical system, including the optical system of the human eye.

Hue an attribute of an object indicating the degree to which it appears similar to one, or as a proportion of two, of the colors red, yellow, green, and blue.

Hue angle a metric of hue which represents the angle of the vector formed between the coordinates of white and the target color.

Illuminance the total luminous flux (c. light) which strikes a surface per unit area.

iLED inorganic Light Emitting Diode

Inorganic Light Emitting Diode a doped inorganic semiconductor device having two electrodes which emits light when a forward current is provided across the electrodes.

LED Light Emitting Diode

LCD Liquid Crystal Display

Liquid Crystal Display a display technology which utilizes materials containing molecules which change their alignment in response to an electric field to modulate the amount of light energy as it passes from a light source to the human eye, permitting the creation of elements that emit different amounts of light. Typically, liquid crystal displays contain a liquid crystal layer between a pair of polarizers. With one alignment, light which passes through one polarizer is absorbed by the second polarizer. In an alternate alignment, a portion of the light passing through the first polarizer passes through the second polarizer.

Light-Emitting Diode an electronic device which conducts current in one direction, producing light proportional to current. LED devices can be formed from either inorganic or organic semi-conductor materials.

Light-Emitting Element the smallest addressable unit within a display. For example in a traditional three color display, any of the red, green, or blue areas comprise an individual light-emitting element.

Lightness the perception elicited by the luminance of an object in a visual environment relative to the perception of the luminance of a perfectly white diffuser in the same visual environment.

Metamerism The ability for a user to perceive two colors having different spectral composition as the same color.

Metamerism Failure The failure of two colors, each having a different spectral composition, to appear equivalent even though they were predicted to appear equivalent.

Microcavity a structure formed by forming two reflecting faces on the two opposing sides of an optical medium. In OLED devices, the reflecting surfaces can be formed by forming the two electrodes, at least partially, from a reflective metal and using the OLED materials themselves to form the optical cavity.

Modulate to vary the strength of a signal. In displays, the term modulate is applied to discuss the ability to vary the strength or the light output from a display device.

Modulation Transfer Function a function characterizing the amount of modulation (c. contrast) a system passes as a function of spatial frequency.

MTF Modulation Transfer Function (MTF)

Multi-Primary Display any display having more than 3 colors of light-emitting elements where at least some portion of the colors produced by the display can be produced by light output from multiple combinations of light-emitting elements. For example, in a multi-primary display including red, green, blue and white light emitting elements a pale yellow can be formed by combinations of light output from red, green, and blue light-emitting elements; light output from red, green, and white light-emitting elements; as well as light output from multiple combinations of red, green, blue and white light-emitting elements.

Optical Infinity the range of distances from a lens where the rays of light reflected from the object and entering the optical system are considered to be parallel to one another.

Observer Metamerism The ability for two observers to perceive two colors having different spectral composition as the same color.

OLED organic light-emitting diode

Organic Light-Emitting Diode a light-emitting diode formed by coating organic (i.e., carbon-based) molecules between an anode and cathode to form a diode. Typical OLEDs are formed from between one and a few thousand angstroms of three or more layers of organic materials where these layers include an electron transport layer to move electrons away from the cathode, a hole transport layer to transport holes from the anode, and a light-emitting layer in which electrons and holes recombine, eventually resulting in the release of light. The number of photons produced is proportional to the number of electrons and holes which recombine.

Pixel in an image file, a pixel is a full color description of a single point within an image. In a display, various combinations of light-emitting elements can be used to render the information associated with a pixel.

Primary in electronic displays, a primary is the color of a light-emitting element capable of providing one of the primary colors; i.e., red, green or blue. In the current text, the term primary represents the color of light produced by any individual light-emitting element within the display.

Relative Luminance the luminance of light reflected by an object in relation to the adapting luminance of the human eye. Many traditional color metrics assume the eye adapts to the luminance of an object with diffuse reflection which reflects 100% of the ambient light across the entire visible spectrum. Therefore, relative luminance can be thought of as the luminance of the object relative to an object having a perfectly diffuse object which reflects all light within the visible spectrum.

Resolution a measure of an imaging system or system component's ability to render detail. In a traditional RGB display, the resolution of the display is typically specified as the inverse of the pixel size where each pixel includes a red, a green, and a blue light-emitting element. This term becomes difficult to specify for display employing color subsampling or for multi-primary displays as the resolution of the display can vary for different colors of light.

Retina the layer of cells on the inside surface of the back of the eye which includes light sensitive and collector cells which connect to the optic nerve.

RGB Displays displays having only red, green, and blue primaries, providing one combination of these three primaries to form each color within the display gamut.

Saturation a measure of the purity of color, usually specified in terms of the distance of a color from the white point of the display within a chromaticity diagram.

Shutter a cover or light block which prevents light to pass to the sensor when it is closed. Typically a shutter is closed in a digital imaging system to provide time to read an image out of a sensor.

Shutter Time the time that the shutter is open in an imaging system permitting light to stimulate the sensor. Increasing the shutter time permits more light to reach the sensor but also increases the likelihood of motion blur as objects in the scene or the camera moves.

Simulator Sickness see Cyber Sickness

Spectral Locus the boundary of perceptible colors depicted in a chromaticity diagram. This boundary represents the chromaticity coordinates of 1 nm bandwidth emitters plotted for all wavelengths within the visible spectrum.

Tricolor display a display having light-emitting elements which emit one of three colors, typically red, green and blue in color.

Vection the perception of self-motion which is stimulated by the visual system.

Vergence the rotation of the user's eyes to center an object of interest on the fovea within each eye.

Virtual Reality the perception that one is in an alternate location than they are physically located. In this text, VR displays are displays which are head mounted and viewed through optics to fill a large portion of the user's visual field.

VR Virtual Reality (VR)

Index

A

Absolute luminance, 10, 23, 24, 69, 76, 83, 121, 218–220
Accommodation, 14, 79, 221–223, 226
Accommodative-vergence conflict, 222
Achromatic, 20
Adaptation, 7, 14, 16–19, 34, 35, 65, 69, 119, 138, 211, 219, 220
Additive color, 5
Addressability, 170, 171, 235
Ambient environment, 92, 120, 219
Amorphous silicon (a-Si), 94, 122, 127
Anode, 54, 107, 108, 110, 113, 117
Anti-aliasing, 67, 173, 174
Aperture, 68, 69, 76, 84, 188
Aperture ratio, 105, 180, 189
Augmented Reality (AR), 211–217, 221–223, 226

B

Backlight, 87, 89–92, 94, 97–102, 104, 129, 185, 186, 191, 207, 217, 218, 220, 235
Bayer, 69, 70
Biorhythm, 229
Bit depth, 112, 236
Blackbody, 43, 46
Blackbody locus, 239, 240
Bottom emitting, 118
Brightness, 10, 22, 23, 27, 28, 31, 40, 57, 60–62, 65, 127, 158
Burn in, 121, 122, 124, 127, 130

C

Capacitor, 92, 93, 127, 129
Cathode, 73, 88, 107, 108, 110, 117

Chroma, 35
Chromatic contrast, 20
Chromaticity, 5, 6, 25, 27–30, 34, 43, 52, 53, 58–61, 63, 75, 97–100, 114–117, 119, 139–142, 145–150, 155, 159, 169, 197, 198, 201, 202
Chromaticity diagram, 5, 6, 27–30, 34, 43, 52, 53, 63, 115, 143, 149, 151, 202
Chrominance, 16, 18, 19, 62, 164–166, 170, 173, 178
Circadian rhythms, 45, 200
Color breakup, 104, 152, 153, 184, 203–206, 208, 229, 232
Color filter, 3, 61, 67, 69–72, 90, 92, 98–100, 120, 123, 125, 130, 152, 164, 186, 187, 192–196, 204, 205, 218, 229
Color patterning, 123–125
Complementary Metal Oxide Semiconductor (CMOS), 77
Color rendering index, 32, 33, 47, 48, 56
Complimentary colors, 166
Cones, 15–17, 19, 22, 23, 25, 36, 69, 198, 200, 224
Contrast, 24, 25, 34, 35, 94, 96, 97, 100–102, 104, 107, 121, 129, 132, 158, 164, 165, 175, 179, 180, 203–205, 211, 216–218, 224
Contrast sensitivity, 164, 165, 178, 180
Convergence, 79–81, 221, 222
Cornea, 13, 36
Correlated Color Temperature (CCT), 43, 44, 46, 55
Cyber sickness, 216, 225, 226

© Springer Nature Switzerland AG 2019
M. E. Miller, *Color in Electronic Display Systems*, Series in Display Science and Technology, https://doi.org/10.1007/978-3-030-02834-3

D

Data line, 92, 93, 127
Daylight, 15, 18, 19, 22, 34, 42–45, 47, 48, 56, 57, 72, 74, 141
Depth, 78–83, 221, 222
Depth of field, 14, 19, 68, 79, 80, 221, 222
Diffusor, 27
Digital Imaging and Communications in Medicine (DICOM), 23, 24, 76
Digital Mirror Device (DMD), 152, 205, 207
Diode, 50, 87, 107, 108, 111, 112, 130
Distance, 15, 17, 23, 29–32, 42, 43, 63, 67, 68, 78–83, 167, 170, 174, 215, 221, 222, 224, 227, 234, 235
Dynamic range, 58, 67, 68, 72, 76–78, 83, 113, 158–160, 216, 229, 234

E

Efficacy, 41, 42, 48, 56, 57, 150, 158, 164, 180, 185, 186, 190–197, 229, 231
Efficiency, 41, 42, 52, 54, 56, 57, 87, 92–94, 99, 100, 103, 105, 107, 109, 111, 116–122, 125, 126, 129, 132, 140, 163, 189–193, 195, 196, 199, 229, 231, 232, 234
8K, 171, 235
Electrode, 92, 93, 108, 109, 116–120, 126–129
Electron, 50, 108, 109, 111
Electron Transport Layer (ETL), 116, 117
Emissive display, 27, 107, 189, 217
Encoding, 73–75, 137, 139, 141, 144, 220, 234
European Broadcasting Union (EBU), 75
Eye, 1, 2, 7, 8, 10, 13–24, 31, 34, 39, 40, 42, 46, 53, 54, 67, 69, 71, 79–81, 87, 102, 119, 121, 138, 140, 150, 152, 164–166, 168, 170, 173, 174, 185, 190, 192–194, 199, 203–206, 213, 215, 218, 221–224, 226, 231

F

Fabry-Perot, 116, 118
Field of view, 36, 80, 211, 214, 215, 224, 225
Flare, 18, 19, 102
Flicker, 104, 203
Fluorescent, 46–49, 56, 75, 87, 90, 98, 101, 114
Focal distance, 221
Fovea, 14–16, 36, 79, 80, 221, 224, 225
Foveated imaging, 224
Full Width at Half Maximum amplitude (FWHM), 114, 116, 117

G

Glare, 241
Grassmann's Laws, 25

H

Head-centric, 214
Heads-Up Display (HUD), 212, 222
Helmet-Mounted Display (HMD), 222
High Definition TeleVision (HDTV), 171, 175, 181, 234
High Dynamic Range (HDR), 76, 78, 100, 102, 158, 159, 216, 235, 236
Holes, 107–109, 113, 123, 124
Hole Transport Layer (HTL), 109, 114, 117, 125, 243
Horopter, 221
Hue, 20, 29, 34, 35, 199

I

Illuminance, 40, 41, 43, 65
Image quality, 35, 71, 73, 89, 104, 136, 152, 158, 164, 168, 174, 175, 177–181, 183, 233
Incandescent, 46–49, 56
Indium Tin Oxide (ITO), 93, 108, 116, 117, 119, 128
In-Plane Switching (IPS), 96, 97
International Commission on Illumination (CIE), 5, 6, 23, 26–31, 33, 44, 137, 140, 141, 178, 197, 200
Intrinsically photosensitive retinal ganglion cells (ipRGCs), 15, 200
Iris, 13

J

Just-Noticeable-Differences (JNDs), 177

L

Lambertian, 54, 108, 119
Lateral geniculate nucleus, 20
Lifetime, 47, 109, 121, 122, 126, 129–131, 163, 164, 180, 184, 189, 220, 231, 232
Lightness, 28, 31, 65, 113
Low Temperature Polysilicon (LTPS), 94, 127, 128
Luminance, 7, 8, 10, 16–20, 22–25, 27, 28, 30, 31, 39–42, 45, 51, 53, 54, 58–60, 62–65, 68–72, 75–78, 83, 84, 94–97, 101, 111–113, 118, 119, 121, 122, 126, 128, 135–141, 143, 146–148, 150, 152–160, 164–170, 173–176, 178–180, 183, 187,

190, 191, 193–195, 197, 198, 201–206, 211, 213, 216–220, 226, 227, 229–232, 234–236

M
Magnocellular layers, 20
Melatonin, 19, 45, 200
Mesopic, 23
Metameric failure, 198–200, 203, 207, 208, 232, 234, 236
Metamerism, 10, 183, 197–199, 229
Microcavity, 116–118, 120, 123
Modulation, 25, 54, 94, 97, 165, 179, 180, 189, 204, 205, 224
Modulation Transfer Function (MTF), 179, 180

N
National Television System Committee (NTSC), 73, 74
Nausea, 225

O
Optic chiasm, 20
Optic disk, 13
Organic layer, 108

P
Panchromatic, 71–73, 84
Parvocellular layers, 20
Passband, 59, 60, 120
Perception, 1–3, 6–10, 14–16, 18–20, 23–25, 28, 33, 35, 36, 40, 45, 48, 78, 79, 87, 112, 170, 172, 211, 225, 226, 230
Phase Alternating Line (PAL), 73, 74
Phosphor, 48, 54–56, 98, 140, 141, 190
Photon, 48, 50, 55, 108
Photopic, 19, 22, 23, 25
Pixel arrangement, 177, 181
Pixel array, 163
Pixel pattern, 175–177, 179, 188, 215
Planckian locus, 43, 44
Polarizer, 91, 92, 94, 120, 121
Power consumption, 10, 57, 91, 97, 101, 102, 104, 123, 126, 130, 136, 149, 164, 184, 186–197, 203, 207, 208, 211, 213, 215, 216, 220, 229
Pulse Width Modulation (PWM), 112
Pulvinar nucleus, 20
Pupil, 13, 14, 17–19

Q
Quad HD, 171, 180, 183
Quantum dot, 55, 56, 99, 100, 234

R
REC709, 75, 141
Reflectance, 27, 34, 41, 57, 119–121
Relative luminance, 6, 7, 10, 59–61, 65, 112, 142–144, 153–155, 193, 194, 196
Relative sensitivity, 15, 16, 21
Render, 75, 83, 135, 136, 138, 142, 144, 156, 159, 166, 168, 173, 174, 176, 177, 179, 180, 220, 222–224, 226, 236
Resolution, 3, 15, 16, 18, 36, 70–73, 78, 83, 88, 93, 94, 101, 104, 107, 113, 131, 163, 164, 166, 167, 170–172, 174–177, 180, 183, 207, 214, 215, 224–227, 229, 232–236
Response time, 97, 207
Retina, 14, 15, 17, 19, 20, 25, 80, 104, 165, 166, 200, 203, 235
Retinal ganglion cells, 14, 16–18, 24
RGBW, 71, 126, 146, 147, 155–159, 163, 164, 167–169, 171, 172, 175–177, 183–185, 187–191, 204–208, 229, 235
RGBY, 167–169, 171, 172, 175, 192–194, 196
Rods, 15–19, 22, 23, 200

S
Sampled, 70, 83, 174
Saturation, 10, 18, 20, 29, 30, 35, 58–65, 75, 92, 97–100, 104, 116, 123, 129, 130, 153, 154, 157, 186, 194, 198, 199, 218, 219, 230, 232
Scotopic, 22, 23
Select line, 93
Self-motion, 225
Shadow mask, 113, 123–125
Shutter, 68, 69, 72, 73, 76, 84
Simulator, 175–177, 225
Society of Motion Picture and Television Engineers (SMPTE), 75
Spatial CIElab (s-CIELab), 178, 179
Spatial Filter, 178
Spatial frequencies, 67, 165, 180
Spatial resolution, 18, 20, 148, 164, 167, 229
SRGB, 62, 63, 77, 95, 137, 142
Standard daylight, 43–45
Stereoscopic, 20, 78–83, 221

Sublimation temperature, 109, 113
Subsampling, 169, 232
Subtractive color, 4
Superior colliculus, 20

T
Tandem OLED, 126
Temporally-modulated, 102
Thin Film Transistor (TFT), 92–94, 122, 127, 129
Threshold voltage, 50, 110, 111
Tone scale, 94–96, 113, 140, 159, 218, 220
Top emitting, 129
Tristimulus, 25–27, 34, 69, 137, 138, 141, 142, 145
Twisted-pneumatic (TN), 97

U
Uniform color space, 29, 31, 32, 34

V
Vapor deposition, 109, 110
Vergence, 79, 221, 226
Viewing angle, 95, 96, 103, 104, 118, 119, 132, 207
Viewing distance, 165, 170–172, 174, 175, 177, 180, 215, 234, 235
Virtual Reality (VR), 211–214, 216–219, 221–223, 225, 226, 235
Visible spectrum, 23, 42, 44, 48, 52, 55, 56, 58–60, 69, 99, 113, 192, 194, 195, 231
Visual angle, 164, 165, 170, 172, 177, 180, 204, 214, 215, 224
Visual cortex, 19, 20
Visual Difference Predictor (VDP), 178
Von kries, 138, 139

W
Weber's Law, 23
White Mixing Ratio (WMR), 147, 149
White point, 30, 31, 35, 58, 63, 74, 75, 97, 135, 136, 138, 139, 141–146, 148–150, 154–157, 168, 169, 188, 193–195, 201, 202, 218–220
World-centric, 214

Printed by Printforce, the Netherlands